JN411377

처음 배우는 군사영어

Building the Base for Military English

처음 배우는 군사영어
Building the Base for Military English
허구성 · 장상혁 · 이강준·선슬기·김채령 공저

정가 15,000원
ISBN 978-89-5568-530-5
발행일 2024년 12월 31일

펴낸곳 양서각
펴낸이 신대영
주소 경기도 가평군 청평면 소돌말길 7-6
전화번호 010-2063-5489
디자인 보라

반품처 서연물류
주소 경기도 파주시 장터고개길 120 창고3
전화번호 010-9228-7723

Contents

01

Army and Profession

By the end of this chapter, you will be able to

- explain the Warrior Ethos and Soldier's creed.
- define the seven Army Values.
- understand Army's role and its mission.

"To be prepared for war is one of the most effectual means of preserving peace."

- George Washington

Part 1. Preview

A. Introduction

The primary responsibility of our Army is to conduct prompt and sustained land combat as part of the joint force. The Army must deliver ready, trained, and equipped forces to meet the demands placed upon it. Our Nation expects the Army to be lethal, agile, and adaptive. As a member of Army, every Soldier must know the Soldier's Creed[01] and live the Warrior Ethos[02]. As a future leader of the Army, you must embody high professional standards and reflect nation's values. The Warrior Ethos demands a commitment[03] on the part of all Soldiers to stand prepared and confident to accomplish their assigned tasks and face all challenges, including enemy resistance[04] — anytime, anywhere.

01 creed : 신조, 신념
02 ethos : 기풍, 풍조
03 commitment : 의무, 책무
04 resistance : 저항

B. Key Doctrinal Terms & Definition

creed 신조	A set of beliefs or aims which guide someone's actions.
joint force 합동군	A force composed of elements, assigned or attached, of two or more Military Departments operating under a single joint force commander.
seize 탈취하다	A tactical mission task in which a unit takes possession of a designated area using overwhelming force.
occupy 점령하다	A tactical mission task in which a unit moves into an area to control it without enemy opposition.
ethos 기풍	The characteristic spirit of a culture, era, or community as manifested in its beliefs and aspirations.
character 인격, 인품	The mental and moral qualities distinctive to an individual.

landpower 지상전력	The ability—by threat, force, or occupation—to gain, sustain, and exploit control over land, resources, and people.
commitment 의무, 책무	The state or quality of being dedicated to a cause, activity, etc
integrity 정직, 고결	The quality of being honest and having strong moral principles; moral uprightness.
unified action 통합행동	The synchronization, coordination, and/or integration of the activities of governmental and nongovernmental entities with military operations to achieve unity of effort.

Part 2. Reading

A. The Army and its mission

The Army, a long-trusted institution, exists to serve the Nation. The Army's mission is to fight and win the Nation's wars through prompt[05] and sustained land combat, as part of the joint force[06]. The Army's contribution to joint operations is landpower[07], which is the ability to promptly gain, sustain[08], and exploit[09] control over land, resources, and people. When necessary, the Army conducts large-scale ground combat operations on land to achieve strategic objectives and impose our will on enemies.

The Army's primary mission is to organize[10] , train, and equip its forces to conduct prompt and sustained land combat to defeat enemy ground forces and seize[11], occupy[12], and defend land areas. It supports four strategic roles for the joint force. Army forces shape operational environments, counter aggression on land during crisis, prevail during large-scale ground combat, and consolidate gains. Army forces accomplish their missions through the operational concept called unified land operations[13]. Unified land operations is the simultaneous[14] execution of offense, defense, stability, and defense support of civil authorities across multiple domains[15] to shape operational environments, prevent conflict, prevail in large scale ground combat, and consolidate gains as part of unified action[16].

05 prompt : 신속한, 즉각적인
06 joint force : 합동군
07 landpower : 지상전력
08 sustain : 지속하다
09 exploit : 이용하다, 전과확대하다
10 organize : 조직하다
11 seize : 탈취하다
12 occupy : 점령하다
13 unified land operations : 통합지상작전
14 simultaneous : 동시의
15 domain : 영역
16 unified action : 통합행동

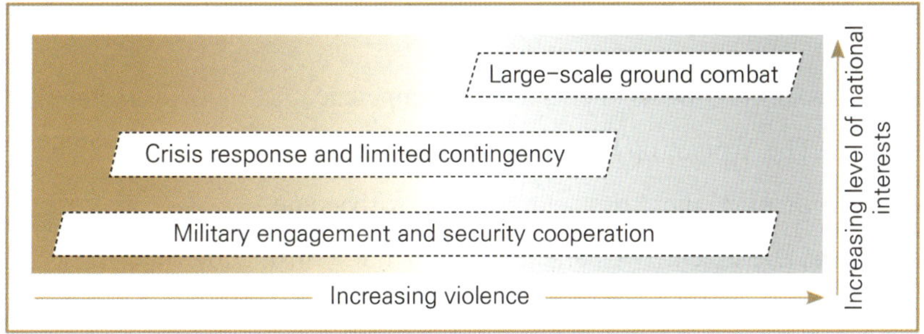

<Operational categories and the spectrum of violence>

B. The Army Values

The Army is a values-based organization. It upholds principles that are grounded in the Constitution and inspire guiding values and standards for its members. These principles are best expressed by the Army Values, Soldier's Creed, and Warrior Ethos. The Soldier's Creed captures the spirit of being a Soldier and the dedication[17] Soldiers feel to something greater than themselves. Embedded in the Soldier's Creed is the Warrior Ethos—the very essence of what it means to be a Soldier. It proclaims the selfless commitment to the Nation, mission, unit, and fellow Soldiers that all Soldiers espouse[18]. The seven Army Values of Loyalty, Duty, Respect, Selfless Service, Honor, Integrity[19], and Personal Courage (LDRSHIP) are universal and they enable you to see what is right or wrong in any situation.

17 dedication : 헌신
18 espouse : 옹호하다, 지지하다
19 integrity : 정직, 고결

<The building block of Army values>

C. The Army in the Profession of Arms

Soldiers are trusted professionals of character[20], competence[21], and commitment. The Army is an honored profession, founded on the bedrock[22] of trust. the Army requires you to become a professional and the essential Characteristics of the Army profession are: Trust, Military Expertise, Honorable Service, Stewardship[23] and Esprit de Corps[24].

20 character : 인격, 인품
21 competence : 자격, 능력
22 bedrock : 기초, 근간
23 Stewardship : 책임관리
24 Esprit de Corps : 단결심, 소속감

• Trust

Trust is the foundation of the Army's relationship with the its people, who rely on the Army to ethically, effectively, and efficiently serve the Nation. Within the Army profession, trust is shared confidence among commanders, subordinates, and partners in that all can be relied on and all are competent in performing their assigned tasks.

• Honorable service

Honorable service is support and defense of the Constitution, nation's people, and the national interest in a manner consistent with the Army ethic. Army professionals serve honorably by obeying the laws of the Nation and all legal orders. Army forces reject and report illegal, unethical, or immoral orders or actions.

• Military expertise

Military expertise is the ethical design, generation, support, and application of landpower, and the associated capabilities essential to accomplishing missions.

• Stewardship

Stewardship is the responsibility of Soldiers and Army

civilians to strengthen the Army as a profession. It includes caring for the people and resources entrusted to them by the nation's people, ensuring Army forces are ready, now and in the future, to accomplish the Army’s missions.

Esprit de corps

Esprit de corps denotes the Army’s winning spirit—a collective ethos of camaraderie, mutual trust, and cohesive teamwork.

The Army’s essential characteristics of trust, honorable service, military expertise, stewardship, and ésprit de corps enable the Army to serve its nation faithfully as an established military profession. These characteristics of the Army Profession reflect our national ideals, the Army Values, the Army Ethic, and the Army’s approach to accomplishing its mission to defend the Constitution and its people.

You exhibit professionalism as an Army officer and leader when you respect the Constitution and the military’s civilian leadership and when you live the Soldier’s Creed and Army Values in all of your daily activities, no matter what your rank, your current job, or where you are assigned.

<The Army profession>

Part 3. Dialogue

[Dialogue 1] A platoon leader and his soldier are talking about Army and its mission.

PLATOON LEADER : Private Choi, do you know why the Army exists?

SOLDIER : Sir, to serve the Nation and fight and win its wars, right?

PLATOON LEADER: Private Choi, do you know why the Army exists?

SOLDIER: Sir, to serve the Nation and fight and win its wars, right?

PLATOON LEADER: Exactly. And how do we achieve that?

SOLDIER: By conducting prompt and sustained land combat as part of the joint force, Sir.

PLATOON LEADER: Correct. Our contribution to joint operations is landpower. Can you explain what that means?

SOLDIER: Sir, landpower is about gaining, sustaining, and exploiting control over land, resources, and people during operations.

PLATOON LEADER: Well said. The Army's mission is established by Congress. Do you remember the key components of that mission?

SOLDIER: Yes, Sir. Preserving peace and security, defending the United States and its territories, supporting national policies, implementing national objectives, and overcoming threats to our peace and security.

PLATOON LEADER: Outstanding. Keep this in mind—every mission we undertake aligns with these objectives.

[Dialogue 2] Two Young Officers Discussing Army Values and Leadership.

2LT PARK: Hey, Lieutenant Jason, we're about to lead Soldiers who look up to us. What do you think is the foundation of being a great leader?

1LT JASON: For me, it starts with the Army Values—Loyalty, Duty, Respect, Selfless Service, Honor, Integrity, and Personal Courage. If we live by those, we set the right example.

2LT PARK: I agree. The Soldier's Creed also comes to mind—it reminds us of our commitment to the Nation, the mission, and our fellow Soldiers.

1LT JASON: Absolutely. The Warrior Ethos in the Soldier's Creed is key. "I will never accept defeat" and "I will never leave a fallen comrade" always stand out to me.

2LT PARK: Right. And don't forget professionalism. Being a trusted professional means we embody those values in everything we do, no matter our rank or position.

1LT JASON: Exactly. Trust, Honorable Service, and Esprit de Corps are essential in leading by example.

2LT PARK: Well said. Let's make sure our actions reflect these principles.

Part 4. Quiz

A. Multiple-Choice Questions

01. What is the primary mission of the Army?

a. Conduct air operations

b. Fight and win the Nation's wars

c. Provide humanitarian aid exclusively

d. Manage international diplomacy

02. What is the Army's contribution to joint operations?

a. Sea power

b. Air superiority

c. Landpower

d. Cyber defense

03. Which of the following is NOT one of the Army Values?

a. Loyalty

b. Integrity

c. Compassion

d. Honor

04. Which of these is a core component of the Soldier's Creed?

a. Commitment to civilian policies

b. Dedication to something greater than oneself

c. Strictly following personal goals

d. Avoiding challenging missions

B. True/False Questions

01. The Army's mission is solely focused on combat operations. (T / F)

02. The Warrior Ethos is embedded in the Soldier's Creed. (T / F)

03. Trust and Esprit de Corps are essential characteristics of the Army profession. (T / F)

04. Living the Soldier's Creed and Army Values is expected only of officers, not enlisted Soldiers. (T / F)

C. Fill-in-the-Blank Questions

01. The seven Army Values are Loyalty, Duty, Respect, Selfless Service, Honor, Integrity, and ________.

02. The Army contributes to the joint force through ________, which involves gaining, sustaining, and exploiting control over land, resources, and people.

Army stories

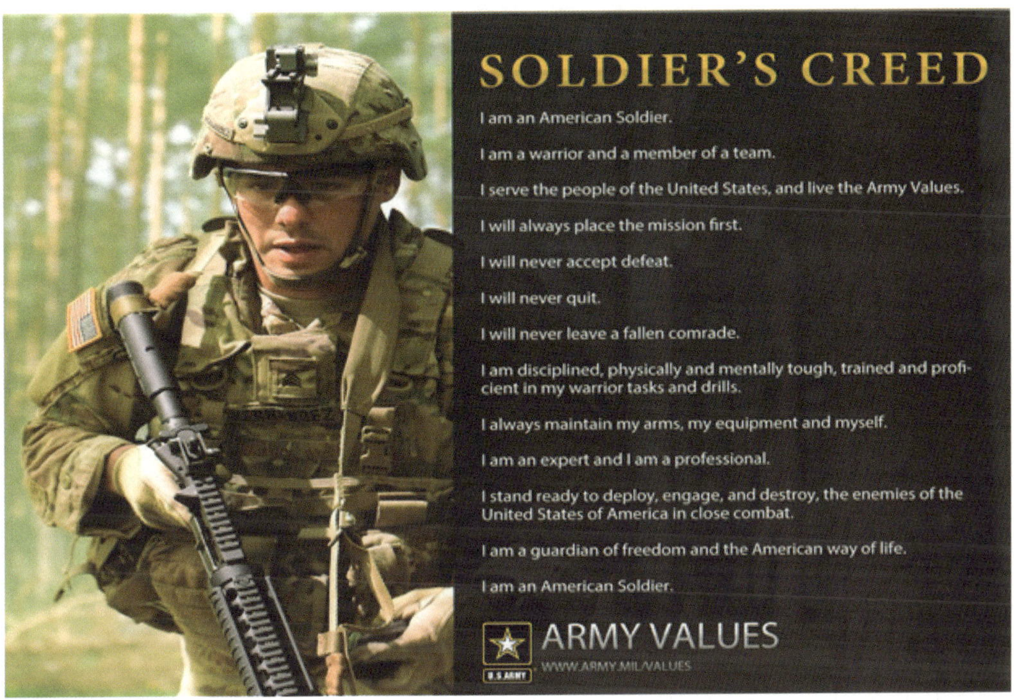

美육군복무신조
나는 미육군 군인이다.
나는 전사요 단체의 일원이다.
나는 미국 국민을 위해 복무하며 군사 덕목을 지킨다.
나는 반드시 맡은 임무를 우선적으로 실시한다.
나는 패배를 인정하지 않는다.
나는 포기하지 않는다.
나는 쓰러진 전우를 버리지 않는다.
나는 몸과 마음을 강하게 하며, 군사 기술에 능통하도록 훈련한다.
나는 항상 무기와, 장비와, 내 자신을 정비한다.
나는 숙련자요 전문가다.
나는 미 합중국의 적을 근접전투에서 교전하고 격퇴하기 위해 파견될 준비가 되었다.
나는 자유와 미국인의 삶의 양식을 수호한다.

나는 미국 군인이다.

대한민국 육 · 해 · 공군은 각각의 복무신조를 가지고 있다. 국군 최초의 복무신조는 1948년 12월 1일에 사병훈을 근거로 제정된 '국군 3대선서'다. 1949년에 '국군맹서'로 개칭된 3대선서는 대통령령 제282호로 군인복무령의 기초가 되었다.

국군 3대 선서 (國軍三大宣誓)	1. 우리는 선열의 혈적을 따라 죽엄으로써 민족국가를 지키자. 2. 우리의 상관, 우리의 전우를 공산당이 죽인 것을 명기(銘記)하자. 3. 우리 군인은 강철같이 단결하여 군기를 엄수하며 국군의 사명을 다하자.
육군 복무신조 (우리의 결의)	우리는 국가와 국민에 충성을 다하는 대한민국 육군이다. 하나. 우리는 자유민주주의를 수호하며 조국통일의 역군이 된다. 둘. 우리는 실전과 같은 훈련으로 지상전의 승리자가 된다. 셋. 우리는 법규를 준수하고 상관의 명령에 복종한다. 넷. 우리는 명예와 신의를 지키며 전우애로 굳게 단결한다.
해군 복무신조 (해군의 다짐)	우리는 영예로운 충무공의 후예이다. 하나. 명령에 죽고사는 해군이 되자. 하나. 책임을 완수하는 해군이 되자. 하나. 전기(戰技)를 갈고닦는 해군이 되자. 하나. 전우애로 뭉쳐진 해군이 되자. 하나. 싸우면 이기는 해군이 되자.
공군 복무신조 (공군의 목표)	대한민국 공군은 항공우주력을 운영하여 첫째, 전쟁을 억제하고 둘째, 영공을 방위하며 셋째, 전쟁에서 승리하고 넷째, 국익증진과 세계평화에 기여한다.

육군의 복무신조는 육군 핵심가치와 일맥상통한다. 육군의 핵심가치란 육군 전 구성원이 추구하여야 할 이념적 방향이며, 참된 군인이자 민주시민으로서의 사고와 행동을 올바르게 결정할 수 있도록 하는 가장 기본이 되는 기준이요 규범이다. 우리 육군은 위국헌신(Selfless commitment), 책임완수(Mission first), 상호존중(Mutual repect)을 핵심가치로 삼고 있다.

육군의 임무는 헌법에 근거하며 국민이 법률에 명시하여 육군에 부여한 과업이다. 국군조직법 제3조1항에서는 "육군은 지상작전을 주 임무로 하고, 이를 위하여 편성되고 장비를 갖추며 필요한 교육훈련을 한다."고 육군의 임무를 명시하고 있다. 이러한 육군의 임무에 기초하여 육군이 역량을 집중해야 할 지향점으로 "육군의 목표"를 제시하고 있으며, 육군의 목표는 국가방위의 중심군으로서 ① 전쟁억제에 기여하고, ② 지상전에서 승리하며, ③ 국민편익을 지원하고, ④ 정예강군을 육성하는 것이라 할 수 있다.

02

Army Ranks and Echelons

By the end of this chapter, you will be able to

- identify the structure and hierarchy of military ranks and units.
- describe the roles and responsibilities of each unit and rank within the military organization.
- understand the relationships between different units and how they work together to achieve mission objectives.

"An army's effectiveness depends on its chain of command. Trust in leadership begins with those at the top leading by example."

- Dwight D. Eisenhower

Part 1. Preview

A. Introduction

The Army profession is a unique vocation[01] of experts who are entrusted to defend the Constitution[02] and the rights and interests of the American people. More than an indication of pay grade[03], Army ranks provide a system of leadership that indicates a Soldier's level of expertise[04], responsibility[05] and authority[06] inside that profession. Regardless of rank, every Soldier has a significant role in the total Army mission. Military rank is more than just who salutes whom. Military rank is a badge of leadership. Responsibility for personnel, equipment and mission grows with each advancement.

01 vocation : 직업
02 Constitution : 헌법
03 pay grade : 급여등급
04 expertise : 전문성
05 responsibility : 책임
06 authority : 권한

B. Key Doctrinal Terms & Definition

Term	Definition
authority 권한	The power or right to give orders, make decisions, and enforce obedience.
chain of command 지휘체계	The succession of commanding officers from a superior to a subordinate through which command is exercised.
echelon 제대	Separate level of command.
squad 분대	A small military unit typically containing two or more fire teams.
platoon 소대	A subdivision of a company or troop consisting of two or more squads or sections.
company 중대	A unit consisting of two or more platoons, usually of the same type, with a headquarters and a limited capacity for self-support.
battalion 대대	A unit consisting of two or more company-, battery-, or troop-size units and a headquarters.

brigade 여단	A unit consisting of two or more battalions and a headquarters company or detachment.
division 사단	An echelon of command and tactical formation that employs brigade combat teams, multi-functional brigades, and functional brigades to achieve objectives on land.
close-combat 근접전투	Warfare carried out on land in a direct-fire fight, supported by direct and indirect fires, and other assets.

Part 2. Reading

A. Army Ranks

Enlisted Ranks[07]

Enlisted Soldiers are the backbone of the Army. They have specific specialties within an Army unit, perform specific job functions and have the knowledge that ensures the success of their unit's current mission within the Army. Enlisted Army ranks are broken down into three groups: Junior Enlisted Ranks(E-1 through E-3), Non-Commissioned Officers(E-4 through E-6), and Senior Non-Commissioned Officers(E-7 through E-9).

07 enlisted rank : 병사와 부사관(NCO)를 포함한 계급

08 Non-Commissioned Officers : 부사관

Junior Enlisted Ranks (E-1 through E-3)

Promotion through E-3 is fairly automatic, while advancement to E-4 requires having two years service and attending a specific training class.

Non-Commissioned Officers (E-4 through E-6)

The United States Army considers all ranks E-4 and above to be Non-Commissioned Officers(NCOs)[08]. Corporals(E-4) are referred to as junior NCOs.

Army Senior Non-Commissioned Officers (E-7 through E-9):

Although the Army does not make the official distinction in the rank structure, enlisted ranks of Sergeant First Class and above(E-7 – E-9) are generally referred to as Senior NCOs.

The Sergeant Major of the Army(SMA) is a rank held by only one enlisted Soldier at a time. The SMA is the senior enlisted advisor to the Army Chief of Staff and is selected based on his or her outstanding leadership, extensive experience, and ability to communicate both up and down the Army chain of command[09].

09 chain of command : 지휘체계(계통)

Insignia	Rank	Paygrade	Insignia	Rank	Paygrade
	Private (PVT/PV2) 이등병	E-2		Sergeant First Class (SFC) 중사	E-7
	Private First Class (PFC) 일등병	E-3		Master Sergeant (MSG) 상사	E-8
	Specialist (SPC) 상등병	E-4		First Sergeant (1SG) 상사	E-8
	Corporal (CPL) 상등병	E-4		Sergeant Major (SGM) 원사	E-9
	Sergeant (SGT) 병장	E-5		Command Sergeant Major (CSM) 주임원사	E-9
	Staff Sergeant (SSG) 하사	E-6		Sergeant Major of The Army 육군주임원사	E-9

* Private (E-1) : 이등병은 계급마크가 없음.

Warrant Officer Ranks

Warrant officers[10] are the adaptive technical expert, combat leader, trainer, and advisor. Through progressive levels of expertise[11] in assignments, training, and education, the warrant officer administers, manages, maintains, operates and integrates systems and equipment across the army operations. Warrant officers are specialists and experts in certain military technologies or capabilities.

10 Warrant officer : 준사관
11 expertise : 전문성

Officer Ranks

The commissioned ranks are the highest in the military. Army officers are called company grade officers[12] in the paygrades of O-1 to O-3, field grade officers[13] in paygrades O-4 to O-6 and general officers in paygrades O-7 and higher. Commissioned officers are the managers, problem solvers, key influencers and planners who lead enlisted Soldiers in all situations. They plan missions, give orders and assign Soldiers tasks.

12 company grade officer : 위관장교

13 field grade officers : 영관장교

Insignia	Rank	Paygrade	Insignia	Rank	Paygrade
	Second Lieutenant (2LT) 소위	O-1		Brigadier General (BG) 준장	O-7
	First Lieutenant (1LT) 중위	O-2		Major General (MG) 소장	O-8
	Captain (CPT) 대위	O-3		Lieutenant General (LTG) 중장	O-9
	Major (MAJ) 소령	O-4		General (GEN) 대장	O-10
	Lieutenant Colonel (LTC) 중령	O-5		General of the Army (GOA) 원수	O-10
	Colonel (COL) 대령	O-6			

* 5성 계급은 전시에만 쓰였음.

<Army Officer Ranks>

B. Echelons

Four soldiers make up a team — a noncommissioned officer and three junior enlisted soldiers. Teams with special functions may also include officers or warrant officers[14]. An infantry fire team might include two riflemen, one of whom is the team leader; a grenadier and an automatic rifleman, who is used when small reconnaissance or special missions are required. Teams can serve as a base-of-fire[15] or as a maneuver[16] element.

14 warrant officer : 준사관
15 base of fire : 화력지원(부대)
16 maneuver : 기동
17 squad : 분대
18 platoon : 소대

Team	Squad	Platoon	Company	Battalion
Ø	●	●●●	I	II
Brigade	Division	Corps	Field Army	Army Group
X	XX	XXX	XXXX	XXXXX

<Army echelons and its symbols>

Squad[17]

Two teams make up a squad, which has 4 to 10 soldiers. In an infantry squad, the teams divide duties: one serves as a base-of-fire element, while the other serves as the maneuver element. A staff sergeant is often in charge.

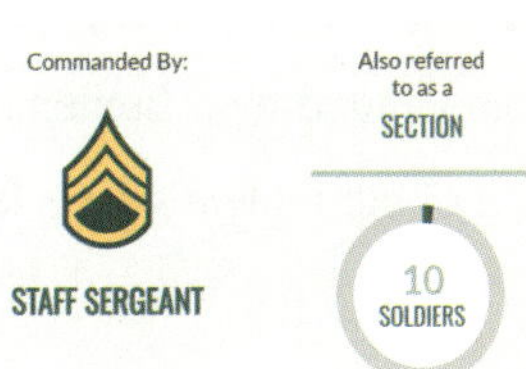

Platoon[18]

A platoon consists of a few squads and up to a few dozen soldiers. They're generally run by a lieutenant, often with a noncommissioned officer as second in command. It is a basic combat unit capable of maneuvering in the conduct of combat operations.An infantry platoon might include a weapons

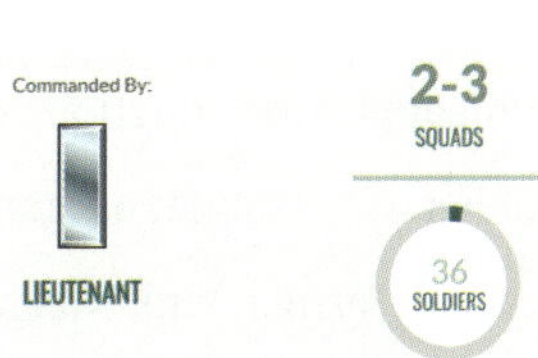

squad[19], which is divided into two M240 machine gun[20] teams, and two close-combat[21] missile teams armed with Javelin missiles.

19 weapons squad : 화기 분대
20 machine gun : 기관총
21 close combat : 근접전투
22 company : 중대

Company[22]

A company has anywhere from a few dozen to 200 soldiers. It's a tactical-sized unit that can perform a battlefield function [23] on its own. A company consists of three or four platoons and is generally commanded by a captain. It can also go by different names, depending on the function. Company-sized artillery units are called batteries, while in armor and air cavalry units, they are called troops. It is capable of receiving and controlling additional combat, combat support or combat service support elements to enhance its mission capability.

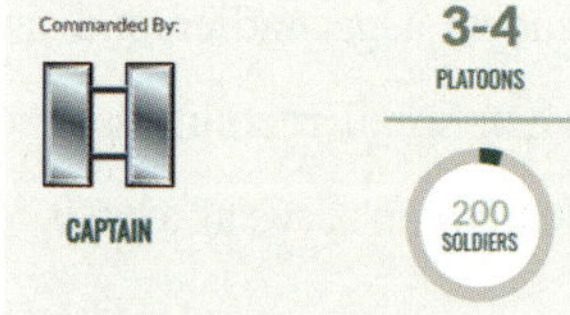

23 battlefield function : 전장기능
24 battalion : 대대
25 brigade : 여단

Battalion[24]

Battalions consist of four to six companies and can include up to about 1,000 soldiers. As a unit that is both tactically and administratively self-sufficient, they can conduct independent operations of limited scope and duration and are usually commanded by a lieutenant colonel. An armor or air cavalry unit of equivalent size is known as a squadron.

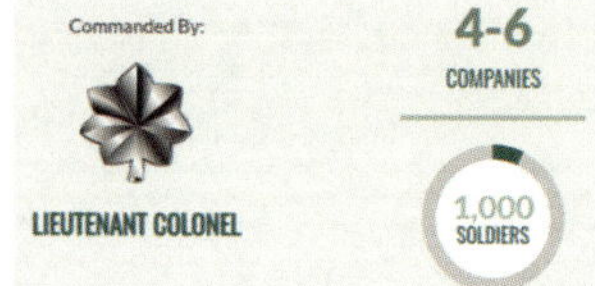

Brigade[25]

A brigade consists of a few battalions that generally include the brigades primary combat arm, as well as sustainment and supporting functions and anywhere from 3,000 to 5,000 soldiers. A colonel is generally in command. For historical reasons, armor and Ranger units of brigade size are called regiments, and the equivalent Special Forces units are called groups. In 2016, the Army reorganized its brigades into brigade

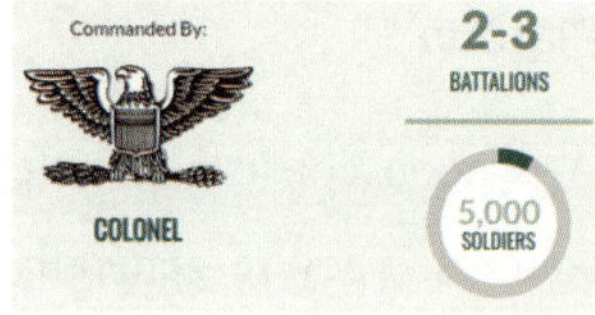

combat teams(BCT), which are autonomous modular brigades that most commonly include one combat arms brigade and its assigned support and fire units.

26 division : 사단
27 corps : 군단

Division[26]

Usually commanded by a major general, divisions are made up of three or four brigades and include 10,000 to 15,000 soldiers. (Divisions are usually comprised of two to three maneuver brigades, a sustainment brigade, and a combat aviation brigade.) Divisions are categorized by one of five types: cavalry, armored, infantry, airborne and artillery. Each can conduct major tactical operations and sustained battlefield operations. They are numbered and assigned missions based on their structures.

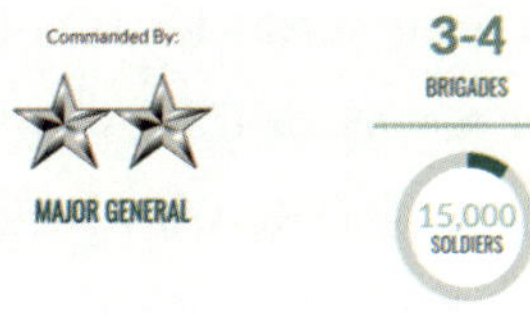

Corps[27]

A corps includes two to five divisions with anywhere between 20,000 and 45,000 soldiers. A lieutenant general is in command. The deployable level of command required to synchronize and sustain combat operations and provides the framework for multinational operations. The corps is the highest level of command that can provide operational direction for actual combat. Higher levels are concerned with administration rather than operations. The current active corps are I Corps at Fort Lewis, Washington; III Corps at Fort Hood, Texas; and XVIII Airborne Corps at Fort Bragg, North Carolina.

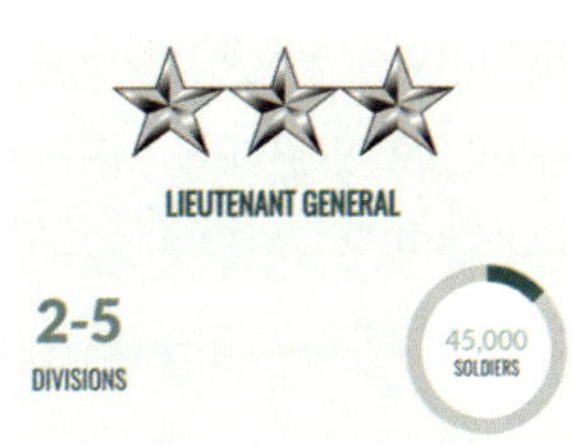

Field Army

A field army consists of two or more corps and is run by a general or lieutenant general. First Army serves as a mobilization, readiness and training command; Third Army, or U.S. Army Central, commands all Army forces for U.S. Central Command; Fifth Army, or U.S. Army North, commands all Army forces for U.S. Northern Command; Sixth Army, or U.S. Army South, commands all Army forces for U.S. Southern Command; Seventh Army, or U.S. Army Europe, commands all Army forces for U.S. European Command; Eighth Army commands all U.S. Army forces in South Korea; and Ninth Army, or U.S. Army Africa, commands all Army forces for U.S. Africa Command.

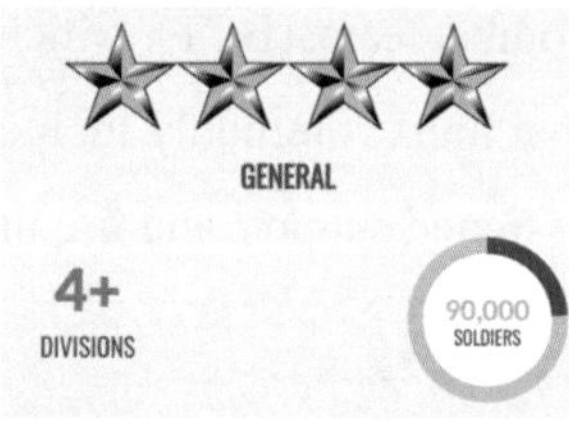

Army Group

A group consists of four or five field armies and between 400,000 and 1 million soldiers. They're commanded by a general and are considered self-sufficient for indefinite periods. They're usually responsible for planning and directing campaigns in particular geographic areas. To differentiate them from field armies, groups are usually written with Arabic numerals (example - 12th Army Group) as opposed to having their number written out.

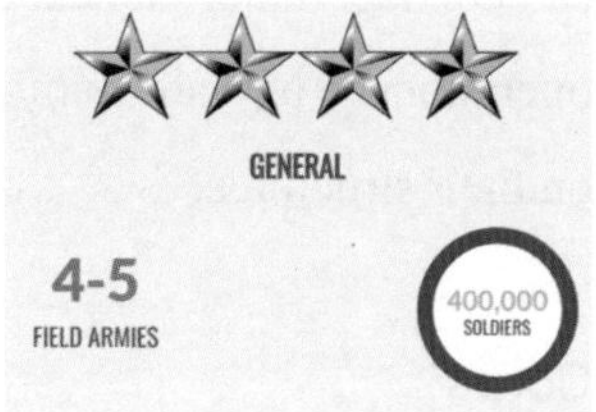

Army Region

These are only used in times of large-scale war, such as World War II. They usually consist of three or more field armies, with about 1 million to 3 million soldiers. They're generally headed by a four- or five-star general.

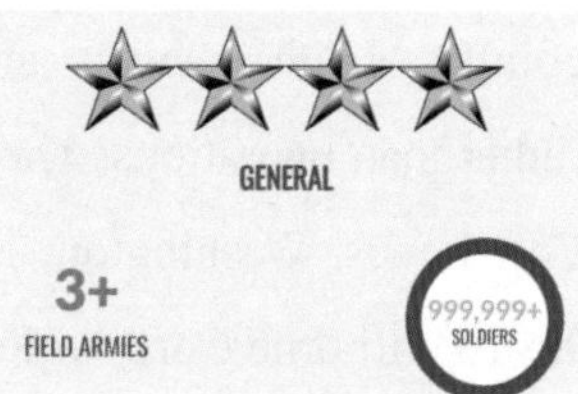

Part 3. Dialogue

[Dialogue] A Lieutenant Explains the Army's Structure to a New Recruit.

LT: Private, do you understand the basic structure of the Army and how it operates?

PVT: Sir, not entirely. I know there are ranks and units, but I'm still confused about how everything fits together.

LT: No problem, let me explain. The Army operates at different levels, starting from the smallest unit, which is the team. A team consists of four Soldiers, typically led by a noncommissioned officer (NCO). For example, a fire team might include a team leader, two riflemen, and an automatic rifleman.

PVT: Got it, sir. So, four Soldiers make a team. What's next?

LT: Two teams make up a squad, which consists of 4 to 10 Soldiers and is led by a staff sergeant. In combat, one team serves as the base-of-fire while the other maneuvers.

PVT: And squads combine to form a platoon, right?

LT: Exactly. A platoon is made up of a few squads, typically led by a lieutenant like myself, with a platoon sergeant as my second in command. Platoons can perform a range of tasks in combat, such as reconnaissance or close combat.

PVT: What about the next level?

LT: A company consists of three or four platoons and is commanded by a captain. Companies are tactical units capable of performing battlefield functions independently.

PVT: That makes sense so far, sir. Does it keep scaling up from there?

LT: It does. Several companies make up a battalion, which is led by a lieutenant colonel. A battalion is self-sufficient and can conduct independent operations for limited durations.

PVT: What's larger than a battalion?

LT: That would be a brigade, consisting of muLTiple battalions and commanded by a colonel.

Brigades are often structured to include their own sustainment and combat support functions.

PVT: Thank you, sir. That really clarified things. One last question—what's the role of enlisted ranks in all of this?

LT: Good question. Enlisted Soldiers like you are the backbone of the Army. You perform specialized tasks and ensure mission success. NCOs provide leadership and training, while officers like me focus on planning and giving orders.

PVT: Understood, sir. I'll do my best to fulfill my role in the team.

LT: That's the spirit, Private. Remember, understanding the structure helps you see where you fit into the bigger picture. Keep studying, and you'll do just fine.

Part 4. Quiz

A. Multiple-Choice Questions

01. Which rank is the leader of a squad?

a. Sergeant Major
b. Staff Sergeant
c. First Lieutenant
d. Corporal

02. What is the primary role of a company in the Army structure?

a. To conduct operations independently for a limited duration
b. To act as a tactical unit capable of performing battlefield functions independently
c. To serve as the deployable level of command for multinational operations
d. To provide operational direction for actual combat

B. True/False Questions

01. A fire team consists of two to three Soldiers and is led by a lieutenant. (T / F)

02. A division typically consists of three to four brigades and is commanded by a major general. (T / F)

C. Fill-in-the-Blank Questions

01. The smallest maneuver element in the Army is the ________.

02. A platoon is typically led by a ________, with a platoon sergeant as second in command.

D. Matching Questions

01. Match the rank to its appropriate description:

1. Commands a company
2. Commands a squad
3. Provides technical expertise in an Army unit
4. Commands a division

a. Staff Sergeant
b. Major General
c. Specialist
d. Captain

E. Scenario-based Questions

01. You are tasked with organizing a reconnaissance mission. Which of the following units would most likely perform this task?

a. Division
b. Fire team
c. Corps
d. Brigade

02. If a lieutenant colonel is commanding a unit, which echelon does this unit most likely belong to?

a. Platoon
b. Battalion
c. Brigade
d. Company

03. During a field exercise, you are assigned to coordinate with a unit responsible for managing 3,500 Soldiers, including sustainment and support functions. This unit is commanded by a colonel. Which type of unit are you working with?

a. Platoon
b. Battalion
c. Brigade
d. Division

Army stories

미 육군의 조직 이해

미국은 전 세계에 걸쳐 11개의 통합전투사령부(UCC, Unified Combatant Command)를 운용하고 있다. 기존의 통합전투사령부는 지역을 담당하는 6개의 지역전투사령부와 4개의 기능전투사령부로 구성되었으나, 2019년 우주사령부가 창설되면서 7개의 지역전투사령부를 운용 중이다. 주한미군이 있는 인도-태평양 사령부 예하에는 태평양 육군, 태평양 공군, 태평양 함대(제3함대, 제7함대), 태평양 해병대, 주일미군, 주한미군, 태평양 특수전 사령부 등이 있다.

Geographic combatant commands 지역사령부			
	Africa Command (USAFRICOM) 아프리카 사령부		Northern Command (USNORTHCOM) 북부 사령부
	Central Command (USCENTCOM) 중부 사령부		Southern Command (USSOUTHCOM) 남부 사령부
	European Command (USEUCOM) 유럽 사령부		Indo-Pacific Command (USINDOPACOM) 인도 태평양 사령부
	Space Command (USSPACECOM) 우주 사령부		
Functional combatant commands 기능사령부			
	Cyber Command (USCYBERCOM) 사이버 사령부		Strategic Command (USSTRATCOM) 전략 사령부
	Special Operations Command (USSOCOM) 특수전 사령부		Transportation Command (USTRANSCOM) 수송 사령부

작전지휘 체계는 대통령에서 국방부 장관, 그리고 각 전투사령부의 전투사령관으로 이어지며, 합동참모본부 의장은 통합전투사령부에 군사적 지휘권은 행사하지 않는다.

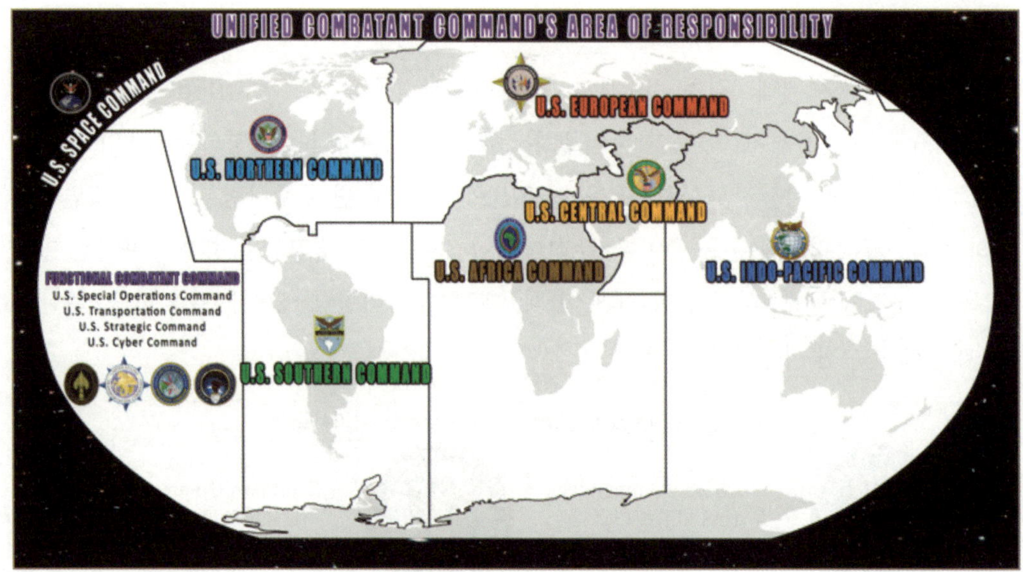

<미 육군 통합전투사령부(출처 : Wikipedia)>

03

Drills

By the end of this chapter, you will be able to

- understand the definition and purpose of drill.
- know how to give the drill command.
- identify individual and unit drills.

"Troops who march in an irregular and disorderly manner are always in great danger of being defeated."

– De Re Militari, Vegetius (A.D. 378)

Part 1. Preview

A. Introduction

Drill is a fundamental aspect of military training, designed to ensure the orderly[01] movement of individuals and units with precision and discipline[02]. It serves multiple purposes, including enabling leaders to efficiently direct their units, instilling[03] habits of immediate response to commands, and fostering leadership skills through the practice of commanding troops[04]. This article provides an overview of the structure and execution of drills, highlighting the roles of drill commands, individual and unit drills, and their importance in maintaining uniformity and order within military operations.

01 orderly : 정연한, 질서있는
02 discipline : 규율
03 instill : 서서히 주입시키다
04 troop : 부대

B. Key Doctrinal Terms & Definition

drill 제식	Certain movements by which a unit (or individuals) is moved in a uniform manner from one formation to another or from one place to another
formation 대형	The arrangement of elements of a unit in a prescribed manner
order 명령	Something that someone tells you you must do
unison 조화	Simultaneous performance or utterance of action or speech
preparatory command 예령	A part of drill command which states the movement to be carried out and mentally prepares the soldier for its execution
command of execution 동령	A part of drill command which tells when the movement is to be carried out
stationary 움직이지 않는	Not moving, or not changing
line 횡대	A formation in which the elements are side-by-side

column 종대	A formation in which the elements are one behind the other
cadence 보속	A uniform rhythm or number of steps or counts per minute
subordinate 예하의	Having a lower or less important position

Part 2. Reading

A. Definition and Purpose

Drill refers to a series of movements by which a unit or individuals are moved in a uniform manner from one formation to another, or from one place to another, with movements executed in unison and with precision. The purpose of drill is to enable commanders or noncommissioned officers (NCOs) to move their unit from one place to another in an orderly manner, to aid in disciplinary training by instilling habits of precision and response to the leader's orders, and to provide for the development of all soldiers in the practice of commanding troops.

05 oral : 구두의

B. Drill Command

A drill command is an oral[05] order of a commander or leader. The precision with which a movement is executed is affected by the way the command is given. Most drill commands have two parts: the preparatory command and the command of execution. Neither part is a command by itself, but the parts are termed commands to simplify instruction. The preparatory command states the movement to be carried out and mentally prepares the soldier for its execution. The command of execution tells when the movement is to be carried out. For example, in the command Forward, MARCH, the preparatory command is Forward and the command of execution is MARCH.

C. Individual Drill

Generally, drill can be divided into two categories – individual and unit drills. Individual drill includes stationary[06] movements, steps and marching, and manual of arms. Stationary movements are executed on current positions without changing locations. Movements are initiated from the Position of Attention[07]. However, some rest movements may be executed from other rest positions[08].

06 stationary : 움직이지 않는
07 position of attention : 차려자세
08 rest position : 쉬어자세

Command	Meaning	Command	Meaning
squad, ATTENTION	차려자세	Left, FACE	좌향 좌
Parade, REST	열중 쉬어	Right, FACE	우향 우
Stand at, EASE	쉬어	About, FACE	뒤로 돌아
AT EASE	쉬어	Present, ARMS	경례
REST	편히 쉬어	Order, ARMS	바로

<Stationary Movements>

All steps and marching movements executed from the Halt[09] are initiated from the Position of Attention. Except for Route Step March and At Ease March, all marching movements are executed while marching at Attention. Marching at Attention is the combination of the Position of Attention and the procedures for the prescribed step executed simultaneously. When executed from the Halt, all steps except Right Step[10] begin with the left foot. Normal marching is executed at quick time[11], a cadence of 120 steps per minute; double-time marching is done at 180 steps per minute.

09 Halt : 제자리에 서, 정지
10 Right Step : 우측 옆걸음
11 quick time : 정상 보속 (=normal cadence)

Command	Meaning	Command	Meaning
Forward, MARCH	앞으로 가	Half Step, MARCH	반걸음으로 가
Change Step, MARCH	걸음 바꿔 가	Mark Time, MARCH	제자리 걸어 가
Rear, MARCH	뒤로 돌아 가	Right (Left) Step, MARCH	우(좌)로 가(옆걸음)
At Ease, MARCH	제걸음으로 가	Backward, MARCH	뒤로 가
Route Step, MARCH	길걸음으로 가	Double Time, MARCH	뛰어 가

<Steps and Marching Movements>

Manual of arms refer to individual drills with weapons[12]. At the Halt, all movements are initiated from Order Arms or Sling Arms, which are the Positions of Attention with the weapons.

12 weapon : 무기

Command	Meaning	Command	Meaning
Order, ARMS	세워 총	Present, ARMS	받들어 총
Parade, REST	열중 쉬어	Inspection, ARMS	검사 총
AT EASE	쉬어	Right Shoulder, ARMS	우로 어깨 총
REST	편히 쉬어	Left Shoulder, ARMS	좌로 어깨 총
Port, ARMS	앞에 총		

<Manual of Arms>

D. Unit Drill

Unit drill can be categorized depending on the unit levels such as squad, platoon and company drills. While performing unit drill as a member of the unit, individual drill movements and the manual of arms are executed. Furthermore, subordinate unit drills are performed during higher unit drills. For instance, while conducting company drill, individual drill movements, manual of arms, and squad and platoon drills are executed.

Command	Meaning	Command	Meaning
FALL IN	모여	Dress Right, DRESS	정식간격 좌우로 나란히
Count, OFF	번호	At Close Interval, Dress Right, DRESS	좁은간격 좌우로 나란히
Close Interval, MARCH	좁은간격	Column Right (Left), MARCH	줄줄이 우(좌)로 가
Normal Interval, MARCH	정식간격	Rear, MARCH	뒤로 돌아가
Double Interval, MARCH	양팔간격	Right (Left) Flank, MARCH	우향(좌향) 앞으로 가
Ready, FRONT	바로	DISMISSED / FALL Out	헤쳐

<Unit Drill>

The squad has two prescribed formations—line and column. However, the squad may be formed into a column of twos[13]

13 column of twos : 2열 종대

from a column formation. When the squad is in line, squad members are numbered from right to left; when in column, from front to rear.

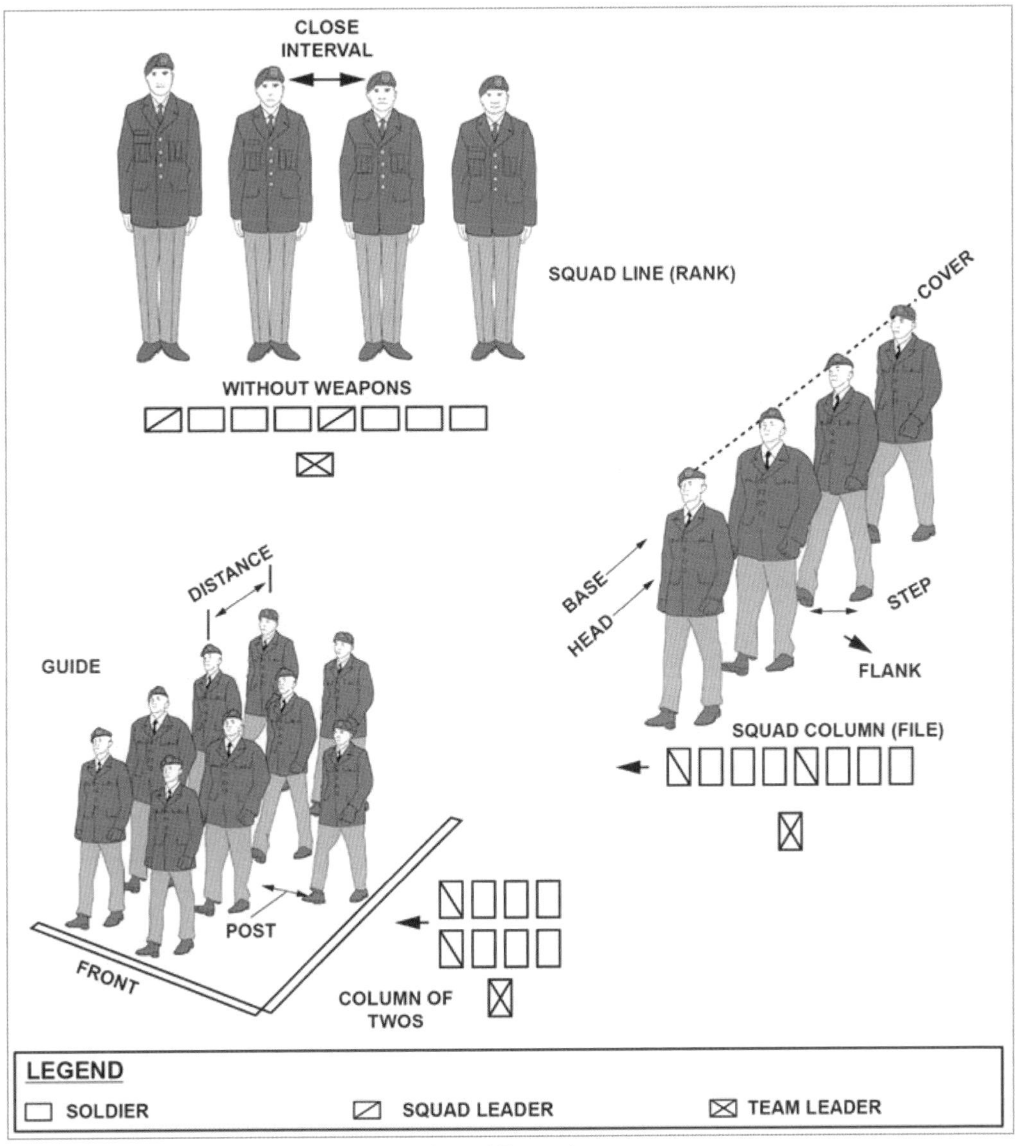

<Squad Formation>

E. Drills and Ceremonies Expressions

Command	Meaning	Command	Meaning
formation	대형	salute to the colors	국기에 대한 경례
alignment	정렬	national flag lowering ceremony	국기 하기식
interval	간격	national flag hoisting ceremony	국기 게양식
double interval	양팔 간격	memorial ceremony	추모식
normal interval	정식 간격	promotion ceremony	진급식
close interval	좁은 간격	retirement ceremony	전역식
distance	거리	wreath-laying ceremony	헌화의식
guide	기준	change of command ceremony	이취임식
head	선두	remarks of outgoing	이임사
rank	오	remarks of incoming	취임사
file	열	line up	도열
line	횡대	moment of silence	묵념
column	종대	gun salute	예포
step	보폭	parade	분열
cadence	보속	parade ground	연병장
quick time	정상보속(분당 120보)	review	열병
double time	속보(분당 180보)	color party	기수단
slow time	완보(분당 60보)	color bearer	기수

Part 3. Dialogue

[Dialogue 1] The platoon is in formation near the flagpole[01] for the national flag hoisting ceremony[02]. Each member has their rifle. The platoon leader gives commands to ensure discipline and precision throughout the event.

PLATOON LEADER: Platoon, ATTENTION!

(The platoon members snap to attention, standing tall with rifles[03] grounded.)

MC: We will now begin the December flag hoisting ceremony. Everyone, please turn around and face the flagpole.

PLATOON LEADER: About, FACE! (The platoon members turn to the right 180 degrees and face the flagpole.)

PLATOON LEADER: Present, ARMS! (After the national anthem[04] ends.)

PLATOON LEADER: Order, ARMS!

MC: Everyone, please turn around and face the front.

PLATOON LEADER: About, FACE! (After all the ceremony ends.)

MC: This ends the December flag hoisting ceremony.

PLATOON LEADER: Platoon, DISMISSED!

01 flagpole : 깃대
02 national flag hoisting ceremony : 국기게양식
03 rifle : 소총
04 national anthem : 국가(國歌)

[Dialogue 2] The dialogue takes place during a morning roll call[05] where the drill instructor ensures all recruits are present and ready for the day.

DRILL INSTRUCTOR: Squad, ATTENTION!

RECRUITS: (Snap to Attention)

DRILL INSTRUCTOR: Report!

SQUAD LEADER: Sir, squad ready for roll call, sir!

DRILL INSTRUCTOR: Fall In!

RECRUITS: (Move quickly into formation, standing at Attention)

DRILL INSTRUCTOR: By the numbers! When your name is called, respond with "Present!"

(Instructor begins roll call)

DRILL INSTRUCTOR: Recruit Kim!

RECRUIT KIM: Present, sir!

DRILL INSTRUCTOR: Recruit Lee!

RECRUIT LEE: Present, sir!

DRILL INSTRUCTOR: Roll call complete. Squad, At Ease!

RECRUITS: (Shift to the At Ease position)

DRILL INSTRUCTOR: Prepare for drill practice. Squad, ATTENTION!

RECRUITS: (Snap back to Attention)

DRILL INSTRUCTOR: Right Face! Forward, MARCH!

---

05 roll call : 점호

Part 4. Quiz

A. Multiple-Choice Questions

01. What is the primary purpose of drill?

a. To entertain soldiers during training

b. To ensure soldiers compete against one another

c. To instill habits of precision and response to orders

d. To provide rest during military exercises

02. Which foot is used to begin all steps from the Halt, except Right Step?

a. Right foot

b. Left foot

c. Either foot

d. Both feet simultaneously

03. What are the two prescribed formations for a squad?

a. Line and column

b. Line and wedge

c. Column and V-shape

d. Wedge and square

B. True/False Questions

01. Drill refers to the movement by which a unit or individuals are moved in a uniform manner from one formation to another, or from one place to another, with movements executed in unison and with precision. (T / F)

02. The command of execution states the movement to be carried out and mentally prepares the Soldier for its execution. (T / F)

03. Manual of arms refers to individual drills with weapons. (T / F)

04. Drill commands consist of a preparatory command and a command of execution, both of which are considered commands by themselves. (T / F)

C. Fill-in-the-Blank Questions

01. The purpose of drill is to enable commanders or noncommissioned officers (NCOs) to move their unit from one place to another in an ______________, to aid in disciplinary training by instilling habits of ______________ and response to the leader's orders, and to provide for the development of all soldiers in the practice of commanding troops.

02. Most drill commands have two parts: the ____________________________ and the ____________________________.

03. While performing unit drill as a unit member, ______________ movements and the manual of arms are executed. Furthermore, ______________ are executed while conducting higher unit drills.

Army stories

Goose Step

Goose step(거위 걸음)은 공식적인 의식이나 분열 등에서 사용되는 독특하고 형식적인 행진 방식으로, 발을 땅에서 높이 들어 올리며 다리를 곧게 유지한 채 행진하는 것이 특징이다. 이는 훈련된 규율과 일사불란한 동작을 상징하는 행진 방식으로 알려져있다. 18세기 프로이센 군대에서 시작된 Goose step은 당시 슈테히슈리트(Stechschritt), 즉 "찌르는 걸음"으로 불리며, 프로이센 군대의 엄격한 훈련과 통제된 움직임을 강조하기 위하여 개발되었다. 이후 독일군을 통해 19세기 러시아로 전파되었고, 20세기에는 소련에 의해 전 세계로 확산되었다. 처음에 Goose step이라는 용어는 느리고 균형 잡힌 행진을 의미했지만, 오늘날에는 주로 경직된 행진 스타일을 가리키는 데 사용된다. 근위병 교대식과 같은 엄숙한 의식에서 주로 사용되지만, 여러 올림픽 개막식에서도 국가와 올림픽 깃발에 경의를 표하기 위해 등장한 적이 있다. Goose step이 나치 독일과 소련 같은 권위주의 정권과 관련되면서 일부 서구사회에서는 이를 권위주의의 상징으로 간주하고 비판하기도 하며, 이 용어를 부정적이고 조롱 섞인 의미로 사용하는 경우가 많다. High step은 Goose step의 변형으로, 다리를 곧게 뻗는 대신 무릎을 꺾어 올리는 방식이며, 일부 군대에서 Goose step를 대체하여 사용된다.

<Goose Step(출처 : Wikipedia)>

<High Step(출처 : Wikipedia)>

Origin of Hand Salute

<Hand Salute(출처 : 육군3사관학교)>

거수경례의 정확한 기원은 확실하지 않지만, 여러 가지 추측이 있다. 첫 번째로, 로마 시대 말기에는 암살이 많았고, 로마 시민들이 공무원을 만날 때 오른손에 무기가 없다는 걸 보여주기 위해 손을 들었다는 설이 있다. 이건 단순히 공격할 의도가 없다는 걸 보여주기 위한 표시였고, 이런 방식이 군대나 민간에서 인사하는 방법으로 발전했다고 한다. 또 다른 설은 중세 기사들에서 유래했다는 것이다. 기사들이 결투 중에 귀족 여성들을 보지 않으려고 손을 들어 눈을 가리거나, 상급자가 다가올 때 손을 들어 투구를 벗어 올리던 동작이 경례의 시작이었다는 얘기다. 하지만 이런 설명에는 반론도 많아서 정확한 기원은 아직 알 수 없다.

거수경례는 시대와 문화에 따라 다양하게 발전했다. 예전에는 두 손을 쓰기도 했고, 검을 들고 다른 손으로 투구를 잡는 방식도 있었다. 현대 군대에서는 좁은 공간에서 경례를 해야 할 때 왼손을 쓰기도 한다. 특히 해군에서는 선박에서 좁은 공간을 지나갈 때 경례하거나, 상급자가 지나갈 때 옆으로 비켜서서 경례하는 '길차렷' 같은 동작을 사용한다.

영국군에서는 1776년 독립전쟁 당시, 미국과 싸우던 영국군 병사들이 모자를 벗고 인사하는 모습을 보여주었고, 이는 후에 군대에서 경례하는 방식으로 정착하게 되었다. 하지만 모자를 벗기 어려운 근위병이나 의장병들은 모자를 벗지 않고 대신 모자의 끝부분을 손으로 잡는 동작을 했고, 이것이 오늘날 군대에서 모자의 끝을 손끝으로 대는 경례로 이어지게 되었다.

거수경례의 기원은 확실히 알 수 없지만, 로마 시대와 중세 기사들에서 유래한 여러 설이 존재하며, 경례의 방식은 시대와 상황에 따라 달라졌고, 오늘날까지도 다양한 형태로 이어지고 있다.

04

Fundamentals of Firing Stance

By the end of this chapter, you will be able to

- explain how to maintain a stable stance and a consistent cheek-to-stock weld for improved accuracy.
- define how to control grip, elbow placement, and trigger use to manage recoil and ensure safety.
- understand how to align sights or aiming devices for accurate targeting and effective follow-up shots.

"Without accuracy, you're just making noise."

- Chris Kyle
(The Author of "American Sniper")

Part 1. Preview

A. Introduction

A well-balanced stance[01] enables a shooter to control recoil[02] and maintain consistent aim. Each element, from hand placement to shoulder positioning[03], plays a crucial role in stabilizing the weapon and improving accuracy. Proper alignment[04] and control allow for smoother, more precise movements, even under challenging conditions. As you learn these core elements, focus on developing muscle memory[05] for each position to make your stance second nature[06]. Mastering these fundamentals is key to enhancing your effectiveness and confidence in the field.

01 well-balanced stance: 균형 잡힌 자세
02 recoil: 반동
03 shoulder positioning: 어깨 위치 잡기(견착)
04 alignment: 정렬, 올바른 맞춤
05 muscle memory: 근육 기억(반복 훈련을 통해 무의식적으로 수행할 수 있는 동작)
06 second nature: 제2의 천성(반복 숙달로 몸에 배어 자연스러운 것)

B. Key Doctrinal Terms & Definition

Term	Definition
stance 자세, 포지션	A shooter's body position and posture used to provide stability and accuracy while firing a weapon.
balance 균형	The shooter's ability to evenly distribute weight to maintain stability and control during aiming and firing.
recoil 반동	The backward movement of a firearm caused by the force of the projectile being fired.
center of gravity 무게중심	The point where a shooter's body weight is evenly distributed, crucial for maintaining stability during shooting.
grip 잡는 법, 그립	The way a shooter holds the firearm to ensure control, stability, and accuracy.
stock weld 개머리판 접촉점; 접용점	The consistent contact between the shooter's cheek and the stock of a rifle, ensuring proper alignment with the sights or scope.

forearm 총의 앞부분; 전완부	The forward part of a firearm's stock, between the receiver and the muzzle, used for support and steadying the weapon during firing.
handguard 총열덮개	A protective covering around a firearm's barrel that shields the user's hand from heat and often provides mounting points for accessories.
trigger 방아쇠	The part of a firearm that is pulled to release the firing mechanism and discharge a round.
trigger guard 방아쇠 울	A protective loop surrounding the trigger to prevent accidental discharge and to protect the trigger mechanism.

Part 2. Reading

A.Building a Stable Foundation – Stance and Grip Basics

A stable stance is very important for accurate shooting[07] because it helps the shooter keep balance and manage recoil after each shot. When a shooter has a steady center of gravity, they can control the force of the shot without losing stability. This balanced stance lets the shooter stay in control, no matter which position they are in, like standing, kneeling, or in a prone position[08]. Each stance can affect how the shooter balances, so it's essential to practice each one carefully.

07 accurate sighting: 정확한 조준

08 prone position: 엎드린 자세

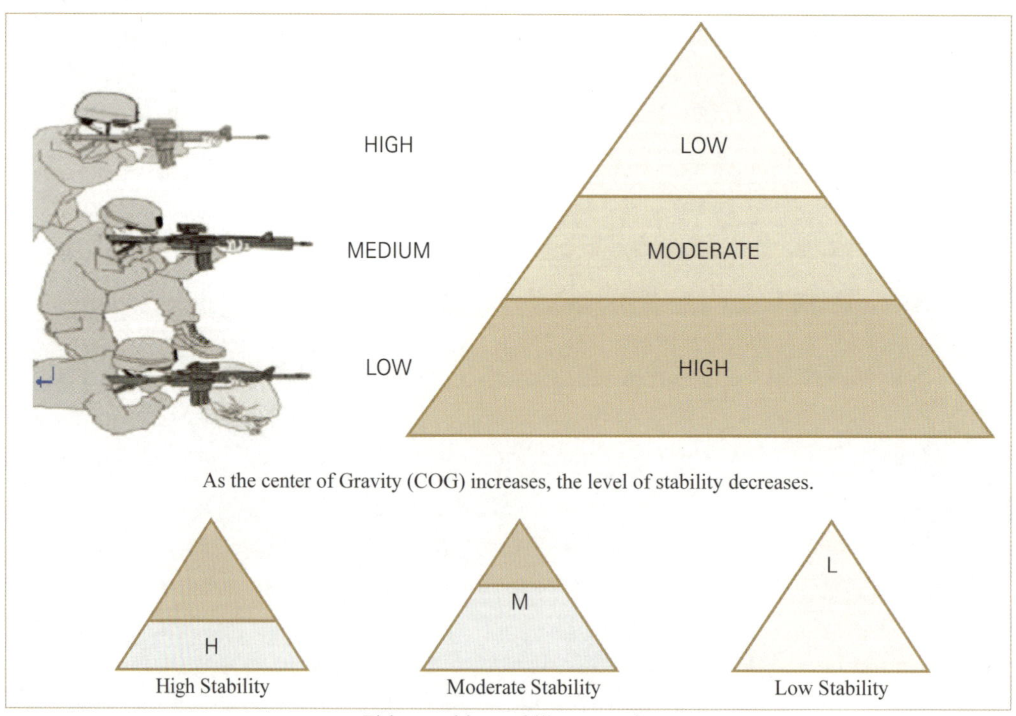

<Firing position stability example>

Establishing a stable stance begins with the feet. For a standing position, the feet should be shoulder-width apart, with the non-dominant foot slightly forward to provide a solid base. This posture allows the shooter to absorb recoil effectively. For kneeling, the dominant knee should be on the ground while the non-dominant leg supports the body, creating a triangular base for added stability. The prone position, where the shooter lies flat with both elbows firmly planted, offers the highest degree of stability and is ideal for long-range precision shooting.

Additionally, practicing transitions between these positions during training can help shooters adapt to various scenarios they might face in the field. For instance, moving smoothly from a standing stance to a prone position can make a significant difference in maintaining accuracy under pressure. Repeated drills focusing on stance stability can help shooters build muscle memory, ensuring they maintain proper alignment even in dynamic environments.

Environmental factors also play a critical role. Uneven terrain, obstacles, or limited visibility can challenge a shooter's ability to maintain a stable stance. Training in diverse conditions helps shooters become adaptable, ensuring they can maintain accuracy whether on soft ground, rocky surfaces, or wet environments. Moreover, understanding how to adjust stances for varying heights and angles, such as firing uphill or downhill, is essential for operational effectiveness.

Having a proper grip on the rifle is also necessary. The nonfiring hand should rest on the forearm near the handguard. This helps keep the rifle steady and supports the weight of the gun. The grip should be firm but relaxed so the shooter can manage the gun without adding too much tension. The placement of the nonfiring hand determines how well the shooter controls the firearm during rapid shots or when changing targets. A forward grip on the handguard provides better leverage for controlling recoil, while a more centered grip can reduce fatigue during extended shooting sessions. For smaller or lighter firearms, shooters might adopt a closer grip to enhance mobility.

09 accidental shots: 실수로 발사된 사격(오격발)

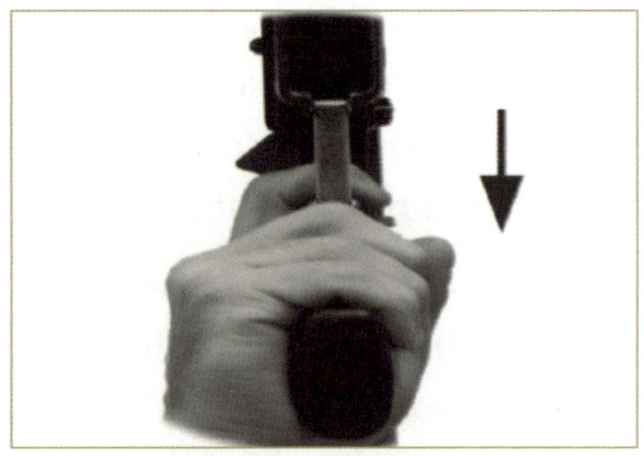

<Natural trigger finger placement>

It's also important to ensure the firing hand maintains consistent contact with the grip while allowing the trigger finger to move freely. If the trigger finger stays outside the trigger guard until it's time to shoot, safety is improved, and accidental shots[09] are avoided. Keeping control of the trigger finger is important for steady shooting, especially in quick situations.

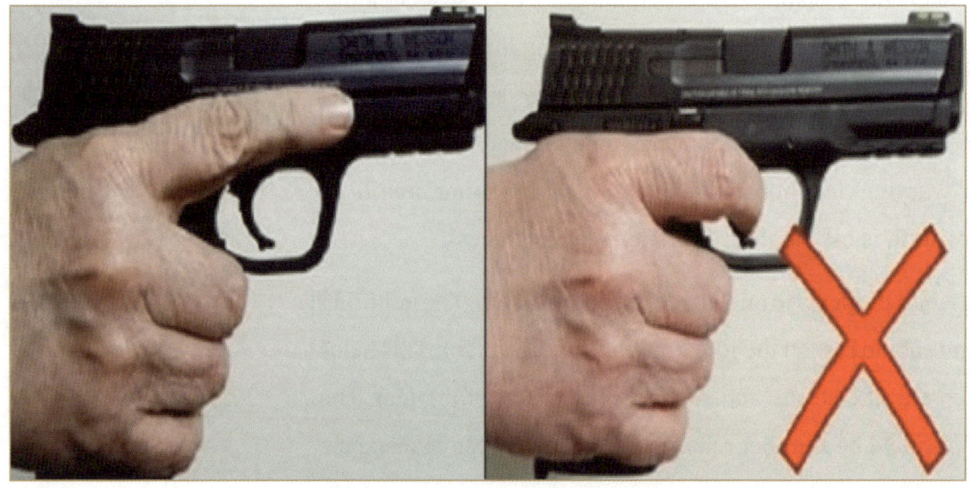

<A proper trigger finger placemen>

Maintaining a relaxed but firm grip on the firing hand prevents unnecessary tension that could pull the rifle off-target. Shooters should practice keeping their trigger finger straight and along the side of the trigger guard until they are ready to fire. This habit not only enhances safety but also improves precision by reducing pre-shot jitters.

Moreover, maintaining a proper grip aids in follow-up shots by minimizing unnecessary movement after firing. This control becomes even more critical when using firearms with significant recoil, as a strong but relaxed grip prevents excessive shifts in the weapon's position. Shooters should regularly evaluate their grip during training to make necessary adjustments and improve overall shooting consistency.

Finally, integrating these principles into regular practice sessions ensures that proper stance and grip become second nature. Combining live fire and dry-firing drills allows shooters to refine their technique and address weaknesses. Practicing with varying firearm weights and sizes helps develop a versatile grip and stance, ensuring readiness for any operational demands. Through consistent effort, shooters can achieve enhanced accuracy, safety, and confidence in their skills.

B. Enhancing Precision – Butt Stock and Alignment

The butt stock[10] should be positioned firmly in the shoulder. This not only helps manage recoil but also keeps the rifle in place, allowing for a good stock weld. The stock weld, where the shooter's cheek meets the butt stock, should be consistent for each shot. This alignment is necessary for accurate sighting.

10 butt stock: 개머리판

The sights must line up properly to make sure that the shot hits the intended target. Good sight alignment means the shooter can focus on aiming without adjusting too much.

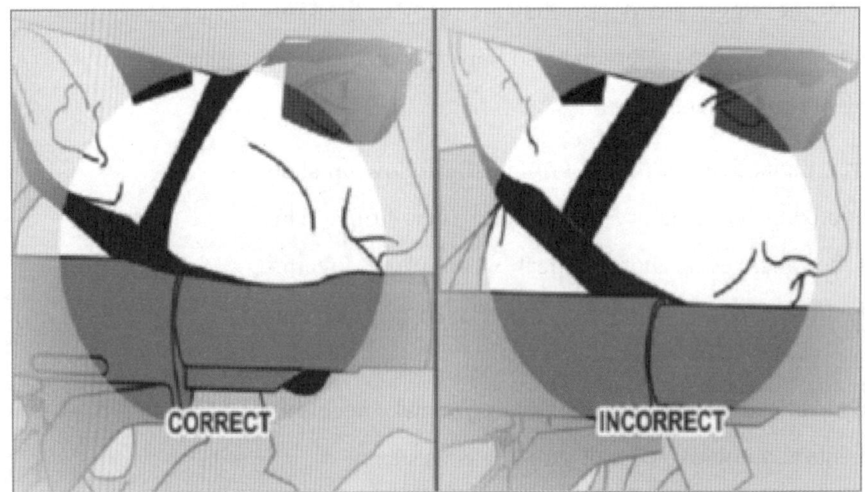

<Stock weld>

The placement of the butt stock not only aids in recoil management but also contributes to overall stability, especially during rapid firing sequences. A firm connection between the shoulder and the butt stock reduces the likelihood of the rifle slipping, ensuring consistent performance even in high-pressure situations. Shooters should practice maintaining this position across various shooting scenarios, from stationary targets to moving ones, to build confidence in their form.

Consistency in stock placement is key to developing accurate and reliable shooting skills. Practicing with the rifle at different angles or positions can help the shooter become familiar with how the stock fits against the shoulder and aligns with the cheek. A well-practiced stock weld reduces unnecessary movement, ensuring the rifle stays on target even after repeated shots.

Developing muscle memory for stock placement is essential. By practicing under different conditions—such as varying

① Support	Unsupported - the Soldier uses his elbows to reduce wobble.
② Leg Position	Spread with heels flat against the ground or the firing side leg bent at the knee to relieve pressure.
③ Stance / Center of Gravity	Prone position maximizes the Soldier's frame against the ground for maximum stability.
④ Firing Elbow	Used to provide stability to the weapon.
⑤ Non-Firing Elbow	As close to directly underneath the weapon as comfortable.
⑥ Firing Hand	Grasping pistol grip, finger off the trigger until ready to fire.
⑦ Non-Firing hand	Firm grasp on the guards.
⑧ Butt Plate	Mid-point of shoulder to absorb recoil impulse.
⑨ Stock Weld	Firm stock weld.
⑩ Shooter-Gun Angle	Shooter body is nearly in-line with the gun-target line.

<Prone, unsupported example>

shooting angles or while wearing body armor—shooters can ensure their alignment remains consistent. A solid stock weld also improves situational awareness, as it allows the shooter to transition between targets smoothly without having to readjust their cheek position repeatedly.

Elbow placement is another key part of the stance. Proper elbow positioning, especially in a prone position, gives extra support to the rifle. By placing the elbow correctly, the shooter creates a stable platform[11] that helps in managing the weight of the rifle and dealing with recoil. This setup makes it easier to stay in control and steady during each shot. The prone

11 stable platform: 안정적인 지지대

position, with the elbow supporting the rifle, is often used for stable, long-distance shots.

In addition to providing stability, correct elbow placement minimizes muscle fatigue, allowing shooters to maintain their stance for extended periods. Practicing elbow placement in positions like the squatting position can prepare shooters for scenarios where prone is not an option. Ensuring that the elbows are positioned symmetrically helps distribute weight evenly, further enhancing the shooter's ability to manage recoil effectively.

① Support	Unsupported-Soldier uses his body for limited weapon support.
② Leg Position	Crouched into a squatting position as low as possible to provide stability.
③ Stance / Center of Gravity	Squatted, leaning toward the target area to assist with recoil management and absorption.
④ Firing Elbow	Resting on the outside of the firing knee using the meaty portion of the tricep for support on the knee.
⑤ Non-Firing Elbow	Resting on the inside of the non-firing thigh (NOT on the knee).
⑥ Firing Hand	Grasping pistol grip, finger off the trigger until ready to fire.
⑦ Non-Firing Hand	Firmly grasping the hand guards or vertical foregrip for support.
⑧ Butt Plate	Mid-point of shoulder to absorb recoil impulse.
⑨ Stock Weld	Firm stock weld.
⑩ Shooter-Gun Angle	Shooter is approximately 30 degrees off the gun-target line.

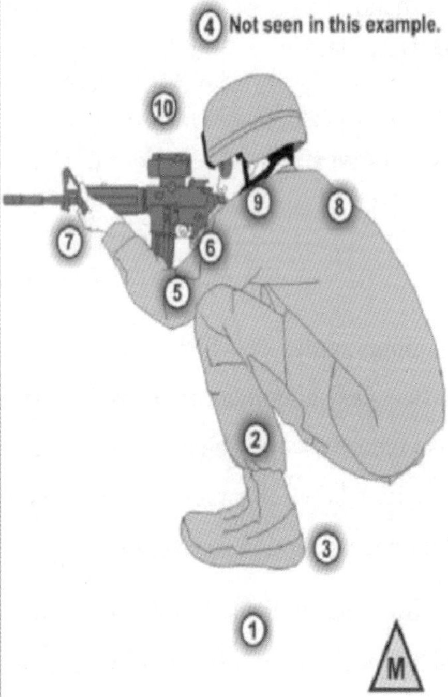

<Squatting position example>

Furthermore, the prone position allows the shooter to minimize their profile and achieve a lower center of gravity, which enhances stability. Practicing elbow placement across different surfaces, such as gravel, grass, or uneven terrain, can help shooters adapt to real-world environments where they may not always have a flat surface to rest on.

The prone position's low profile also makes it ideal for avoiding detection in tactical or combat situations. Training in diverse terrains, including wet or slippery conditions, ensures the shooter can maintain stability regardless of environmental challenges. This adaptability is critical for both precision shooting and operational readiness in dynamic environments.

To further enhance precision, shooters should focus on their breathing patterns. Taking a deep breath, pausing during the natural exhale, and pulling the trigger smoothly ensures minimal rifle movement. Coupled with proper alignment and support from the elbows, this technique can significantly improve accuracy.

Proper breathing control not only steadies the rifle but also helps the shooter remain calm under stress. Incorporating breathing exercises into training routines enhances this skill, ensuring that each shot is deliberate and controlled. Additionally, synchronizing breathing with sight alignment and trigger pull can make the difference between a missed shot and a direct hit.

Lastly, dry-firing drills[12] while in the prone position can reinforce proper alignment and elbow placement. Regular practice under varied conditions, such as low light or simulated stress scenarios, ensures the shooter is prepared for any challenge.

12 dry-firing drills: 비사격 훈련

Dry-firing is particularly useful for improving muscle memory and refining trigger discipline without the distraction of live ammunition. Combining these drills with simulated scenarios, such as time-pressure exercises or shooting at unpredictable targets, prepares the shooter for real-world challenges. By reinforcing these fundamentals, shooters can build a foundation of accuracy, stability, and confidence that will serve them in any situation.

13 aiming device: 조준 장치
14 scope: 조준경
15 iron sights: 철제 조준기 (e.g., 가늠자, 가늠쇠)
16 follow-up shots: 후속 사격(추가 사격)

C. Perfecting Accuracy – Sight Alignment and Habit Formation

The aiming device[13], like a scope[14] or iron sights[15] , must be lined up with the shooter's view. This alignment is important for fire control, helping the shooter maintain accuracy. When all these elements are followed correctly, the shooter's accuracy improves, allowing for more precise follow-up shots[16]. Consistent sight alignment and steady positioning make it possible to hit targets with better control, especially when multiple shots are needed.

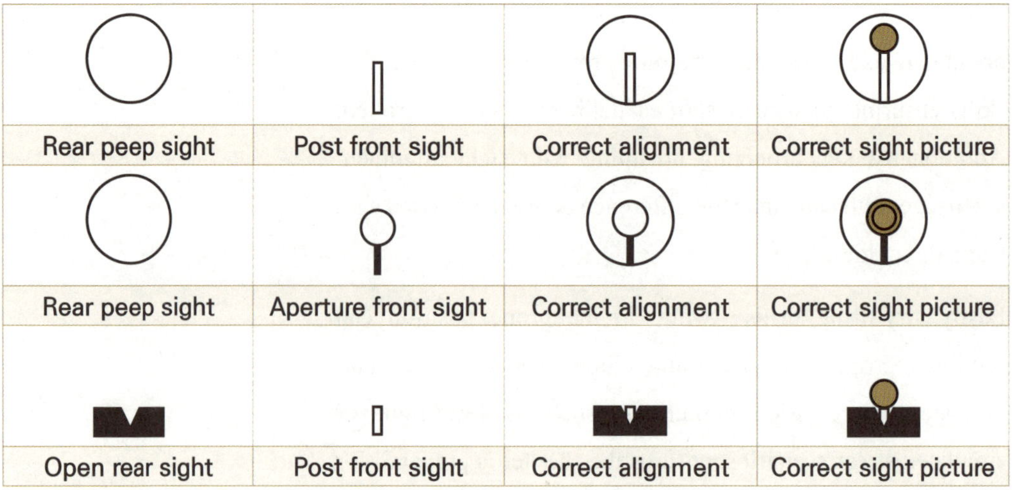

Rear peep sight	Post front sight	Correct alignment	Correct sight picture
Rear peep sight	Aperture front sight	Correct alignment	Correct sight picture
Open rear sight	Post front sight	Correct alignment	Correct sight picture

<Proper Sight Alignment or Sight Picture>

Achieving proper sight alignment requires patience and consistent practice. Shooters should focus on aligning the front and rear sights precisely with the target while maintaining a clear focus on the front sight. For scoped rifles, adjusting the parallax setting and ensuring a proper eye relief distance is crucial to avoid distortion and maintain a clear field of view.

In summary, each part of the stance—stance balance, grip, nonfiring hand placement, trigger control, butt stock positioning, elbow placement, and stock weld—works together to create stability. The goal is to stabilize the rifle, control recoil, and keep a clear line of sight. Practicing these fundamentals helps the shooter develop muscle memory and become more effective and confident in different shooting situations. By focusing on these basics, the shooter can gain better control, accuracy, and comfort in firing, no matter the environment[17].

Regular training sessions incorporating live fire and dry-firing drills help shooters solidify these fundamentals. Deliberate practice under various conditions, such as shooting from unstable platforms or in challenging weather, builds adaptability and confidence. Additionally, incorporating breathing control and trigger discipline enhances the shooter's ability to stay calm and focused, even in high-pressure scenarios.

To further refine accuracy, shooters can use tools such as laser bore sighters to confirm alignment before live firing. Reviewing shot groupings after each session and making necessary adjustments to sights, stance, or grip ensures steady progress. Combining these technical adjustments with consistent practice prepares shooters to perform effectively in any shooting environment.

17 no matter (what) the environment (is / may be) : 어떠한 상황에서도
* 문맥 의존성에 따른 의미전달이 가능하여 간결성 차원에서 완벽한 일부 문법적 요소를 생략하는 경우임.

Part 3. Dialogue

[Dialogue] An instructor teaches a cadet the basics of shooting: stance, grip, trigger control, and sight alignment.

INSTRUCTOR: Let's start with the basics. A comfortable stance is essential to keep balance and absorb recoil. Why is that important?

CADET: I assume it's to keep balanced, right?

INSTRUCTOR: Exactly. Balance helps you control recoil. Your center of gravity should support your weight in different positions—standing, kneeling, or prone.

CADET: Got it. What about my hands?

INSTRUCTOR: Your non-firing hand supports the forearm near the handguard, firm but relaxed. The firing hand controls the trigger with the finger outside the guard until ready.

CADET: (adjusting his grip slightly, looking up for confirmation.) So, my grip should be firm but not too tight?

INSTRUCTOR: (pressing the butt stock firmly into his shoulder and lowers his cheek onto the stock, showing how to align the sights consistently.) Yes. Now, the shoulder. Place the butt stock firmly to manage recoil and keep steady. This also helps with consistent stock weld and sight alignment.

CADET: And my elbow?

INSTRUCTOR: (dropping into a prone position, stabilizing his rifle with his elbow on the ground) Proper elbow placement stabilizes the rifle, especially in prone. Lastly, bring the rifle to your head to keep sight alignment for follow-up shots.

Part 4. Quiz

A. Multiple-Choice Questions

01. Why is having a balanced stance important in shooting?

a. To shoot faster

b. To maintain accurate aim

c. To control recoil

d. All of the above

02. What is the role of the nonfiring hand?

a. To pull the trigger

b. To support the weight of the rifle and provide stability

c. To rest on the trigger guard

d. To hold the stock weld

03. Why should the trigger finger stay outside the trigger guard until ready to shoot?

a. To prevent accidental firing

b. To reduce recoil

c. To improve aim

d. To make the grip firmer

04. What is the purpose of a consistent stock weld?

a. To keep sight alignment steady

b. To support the forearm

c. To manage the recoil

d. To improve firing speed

B. True/False Questions

01. A stable stance helps the shooter keep balance and manage recoil after each shot. (T / F)

02. The center of gravity does not affect the shooter's stability during firing. (T / F)

03. The nonfiring hand should rest on the forearm near the handguard to help keep the rifle steady. (T / F)

04. The grip should be loose to allow for quick movements. (T / F)

05. Keeping the trigger finger outside the trigger guard until ready to shoot improves safety. (T / F)

06. The butt stock should be positioned loosely in the shoulder for flexibility. (T / F)

07. Proper elbow placement, especially in the prone position, helps support the rifle's weight. (T / F)

08. Sight alignment is not necessary for accurate aiming. (T / F)

09. The aiming device must be aligned with the shooter's view to maintain accuracy. (T / F)

10. Practicing a stable stance helps the shooter develop muscle memory for better control and accuracy. (T / F)

C. Fill-in-the-Blank Questions

01. A stable __________ is important for accurate shooting because it helps the shooter keep __________ and manage recoil after each shot.

02. When a shooter has a steady __________, they can control the force of the shot without losing stability.

03. The __________ hand should rest on the __________ near the handguard to help keep the rifle steady.

04. The __________ finger should stay outside the trigger guard until it's time to shoot, which improves __________ and avoids accidental shots.

05. The __________ should be positioned firmly in the shoulder to help manage recoil and allow for a good stock weld.

06. Proper __________ placement, especially in the prone position, provides extra support to the rifle.

Army stories

엎드려 쏘기 자세(prone position)의 기원과 발전

엎드려 쏘기 자세는 오늘날 가장 안정적이고 효과적인 사격 자세로 알려져 있다. 이 자세는 단순히 사격 기술의 한 형태로만 보이지만, 사실 그 기원과 발전 과정은 깊은 역사적, 전술적 의미를 담고 있다. "Prone position"이라는 용어는 라틴어 pronus에서 유래했으며, 이는 "앞으로 기울어진" 혹은 "내려다보는"이라는 뜻을 가진다. 이 자세는 총기의 발전과 함께 그 중요성이 강조되었고, 현대 전술에서도 없어서는 안 될 기본 기술로 자리 잡았다.

초기 총기의 발전과 서서 쏘기 자세의 필요성

화약 무기가 처음 등장했을 때, 사수들은 주로 서서 쏘기 자세를 취했다. 당시 사용된 머스킷과 같은 무거운 화약 무기는 조준 장치가 없었기 때문에 사수는 총을 가슴 높이로 들어 올려 목표물을 겨냥해야 했다. 머스킷의 발사 충격은 상당했기 때문에, 사수는 다리를 넓게 벌리고 무게 중심을 낮춰 반동을 흡수할 수 있는 자세를 유지해야 했다.

서서 쏘기 자세는 사격의 기동성을 제공했지만, 안정성은 부족했다. 이로 인해 초기에는 장거리 사격보다는 근거리 집단 사격이 일반적이었다. 그러나 총기의 발사 기술이 점차 정교해지고, 소총이 등장하면서 장거리 정확도가 중요한 요소로 떠오르기 시작했다. 이에 따라 보다 안정적인 자세가 필요하게 되었고, 엎드려 쏘기 자세가 주목받게 되었다.

19세기: 엎드려 쏘기 자세의 전술적 필요성

엎드려 쏘기 자세는 19세기 전쟁에서 그 중요성이 더욱 부각되었다. 특히 미국 남북전쟁과 유럽의 크림 전쟁에서, 병사들은 개활지에서 적의 사격을 피하기 위해 몸을 낮추는 자세를 채택했다. 엎드린 자세는 적에게 노출되는 면적을 최소화할 뿐만 아니라, 소총의 무게를 팔꿈치로 분산시켜 장거리 사격의 안정성을 극대화할 수 있었다.

당시, 전장에서 저격수의 역할이 커지면서 엎드려 쏘기 자세는 특히 저격술에 필수적인 기술로 자리 잡았다. 미국 남북전쟁에서 활약한 전설적인 저격수 허샤우드 패밀리는 엎드려 쏘기 자세를 활용하여 장거리에서 적의 주요 목표를 정확히 제거한 사례로 유명하다. 이 자세는 전장의 생존율을 높이는 동시에 전술적 우위를 제공하는 중요한 기술로 인정받았다.

제1차 세계대전: 엎드려 쏘기 자세의 표준화

엎드려 쏘기 자세가 군사 교범에 공식적으로 포함된 것은 제1차 세계대전 때부터이다. 이 전쟁에서는 참호전이 주요 전술로 사용되었으며, 병사들은 좁은 참호 안에서 적의 사격을 피하면서도 목표물을 정확히 겨냥해야 했다. 엎드린 자세는 병사들이 몸을 최대한 낮추고, 참호 위로 총을 겨냥할 수 있게 해 주었다.

이 자세는 병사가 긴 시간 동안 안정적으로 사격할 수 있도록 도와주었으며, 총기의 반동을 효과적으로 흡수하는 역할도 했다. 제1차 세계대전 당시 독일의 슈투르무트(돌격부대)는 엎드려 쏘기 자세를 활용하여 참호에서 효과적인 방어와 공격을 수행했다. 이 시기를 기점으로 엎드려 쏘기 자세는 군사 훈련에서 필수적인 기술로 자리 잡게 되었다.

현대 군사 훈련에서의 역할

오늘날 엎드려 쏘기 자세는 미군의 TC 3-22.9 Rifle and Carbine과 같은 훈련 교범에서 가장 기본적이고 중요한 자세로 다뤄진다. 이 교범은 엎드린 자세가 장거리 정밀 사격에서 가장 안정적이며, 다양한 지형에서도 효과적으로 사용할 수 있음을 강조한다.

예를 들어, 병사가 풀밭이나 자갈밭과 같은 불규칙한 지형에서도 엎드려 쏘기 자세를 유지할 수 있도록 훈련한다. 또한, 현대 전술에서는 은폐와 엄폐를 동시에 수행하며, 장시간 안정적으로 목표물을 겨냥할 수 있는 기술로서 이 자세를 채택하고 있다.

결론

엎드려 쏘기 자세는 단순히 안정적인 사격 자세를 넘어서, 병사들의 생존과 정확도를 위한 전술적 선택의 결과물이라 할 수 있다. 이 자세는 전장에서 병사들이 불리한 상황에서도 효과적으로 대응할 수 있도록 도와주었으며, 현대 전쟁에서도 여전히 중요한 역할을 하고 있다.

또한, 이 자세의 역사적 발전은 총기의 발전과 밀접하게 연결되어 있다. 엎드린 자세는 단순히 개인의 기술이 아니라, 전장의 변화에 따른 병사들의 생존 전략으로서 발전해온 것이다.

05

Key Components of Proper Aiming

By the end of this chapter, you will be able to

- understand the critical components of proper aiming.
- learn how to maintain consistency in aiming.
- recognize the relationship between the point of aim and the desired point of impact.

""He who sees first, shoots first, and hits first, wins."

- Colonel John Boyd (1927-1997)

Part 1. Preview

A. Introduction

Maintaining sight alignment between the front sight post and rear sight aperture is essential for a clear line of aim. Consistent stock weld ensures stability and alignment, while focusing on the front sight post sharpens accuracy. Additionally, accounting for bullet drop and making necessary hold-off adjustments are critical for precise targeting at varying distances.

B. Key Doctrinal Terms & Definition

aiming 조준	The process of aligning the firearm's sights with the target to ensure accurate shooting.
marksmanship 사격술	The skill and practice of shooting accurately at a target.
sight alignment 조준선 정렬	The correct positioning of the front and rear sights in relation to the shooter's eye, ensuring they are aligned properly for accurate shooting.
target alignment 표적 정렬	The process of positioning the firearm so that the aligned sights are directed at the center of the target.
sight picture 조준상	The visual image obtained when the aligned sights are centered on the target, indicating proper aim.
bullet trajectory 탄도	The curved path a bullet follows from the firearm's muzzle to the target due to gravity and other factors.
front sight post 가늠쇠	The front component of a firearm's sighting system, typically a vertical post used in conjunction with the rear sight for aiming.
rear sight / rear aperture 가늠자〈경〉	The rear component of a firearm's sighting system, which may feature an open notch or an aperture (peep sight) through which the shooter aligns the front sight.

point of aim 조준점	The exact spot on the target where the shooter intends the bullet to impact.
point of impact 탄착점	The actual location where the bullet strikes the target.
accuracy 정확도	The degree to which a shooter's point of impact matches their point of aim, reflecting how closely the shots align with the intended target.
hold-off 선조준/리드샷	he deliberate aiming adjustment made to compensate for factors like wind or target movement, involving aiming away from the target's center to achieve a hit.

Part 2. Reading

Aiming accurately is a fundamental skill in marksmanship, ensuring that a shooter effectively engages a target with precision[01] and consistency. Proper aiming is built upon several critical components; sight alignment, target alignment (or sight picture), focus on the front sight, and maintaining consistency. Additionally, understanding the relationship between the point of aim and the desired point of impact is crucial for achieving optimal results.

01 precision : 정확성
02 misalignment : 비정렬

A.Sight Alignment

Sight alignment refers to the precise relationship between the front sight post, rear sight aperture, and the shooter's eye. This alignment is essential for accurate shooting, as even minor deviations can result in significant errors. The shooter must center the front sight post within the rear aperture while maintaining an equal amount of light on either side of the post. This alignment creates a straight visual path, ensuring the trajectory of the projectile corresponds closely to the shooter's intent. Importantly, the focus should remain on the front sight post, as it serves as the primary guide for achieving a proper sight alignment.

Achieving and maintaining proper sight alignment is a skill that requires careful attention and repeated practice. Small misalignment[02] may seem negligible but can cause significant deviations in the point of impact, especially over long distances. For example, a slight tilt of the firearm or an uneven light gap between the front sight post and rear aperture can cause shots to land inches off target at 100 meters and much farther at extended ranges.

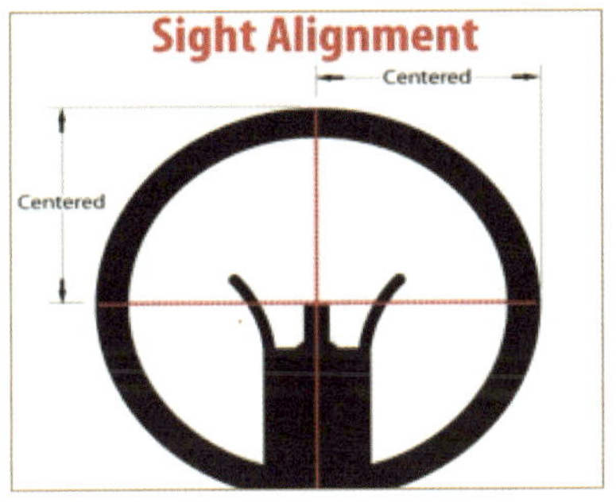

<a correct sight alignment example>

Proper sight alignment also sets the foundation for more advanced aiming techniques. Shooters who master this fundamental can adapt to different sight systems, including red-dot sights[03], holographic optics[04], and magnified scopes[05]. Each of these systems relies on the principle of aligning the aiming reticle[06] or dot with the target, underscoring the universal importance of sight alignment across weapon platforms.

Feature	Red-Dot Sights	Holographic Optics	Magnified Scopes
Magnification	None (1x)	None(1x), works with magnifiers	Variable (e.g., 3-25x)
Reticle Type	Simple dot	Complex holographic reticles	Precise reticles with markings
Field of View	Wide	Wide	Narrow (varies with magnification)
Range	Close to medium (〈200 meters)	Close to medium (〈300 meters)	Medium to long (〉300 meters)
Power Consumption	Low (long battery life)	Moderate (laser technology)	None or battery for illuminated reticles
Best For	Speed and simplicity	Precision and versatility	Long-range precision
Example			

03 red-dot sights : 레드 도트 조준기

04 holographic optics : 홀로그램 조준 장치

05 magnified scopes : 확대 조준경

06 aiming reticle : 조준망

B. Target Alignment (Sight Picture)

Target alignment, commonly referred to as sight picture, involves the placement of the aligned sights on the target. This step combines sight alignment with the visual image of the target through the sights, providing a complete picture of the firing solution. The shooter positions the front sight post, rear aperture, and the target in a single visual plane[07]. When executed properly, the front sight post should appear sharp and in focus, while the target and rear aperture remain slightly blurred. This balance ensures that the shooter maintains accuracy while compensating for the limitations of human vision.

07 visual plane : 시각적 평면
08 situational awareness : 상황인식

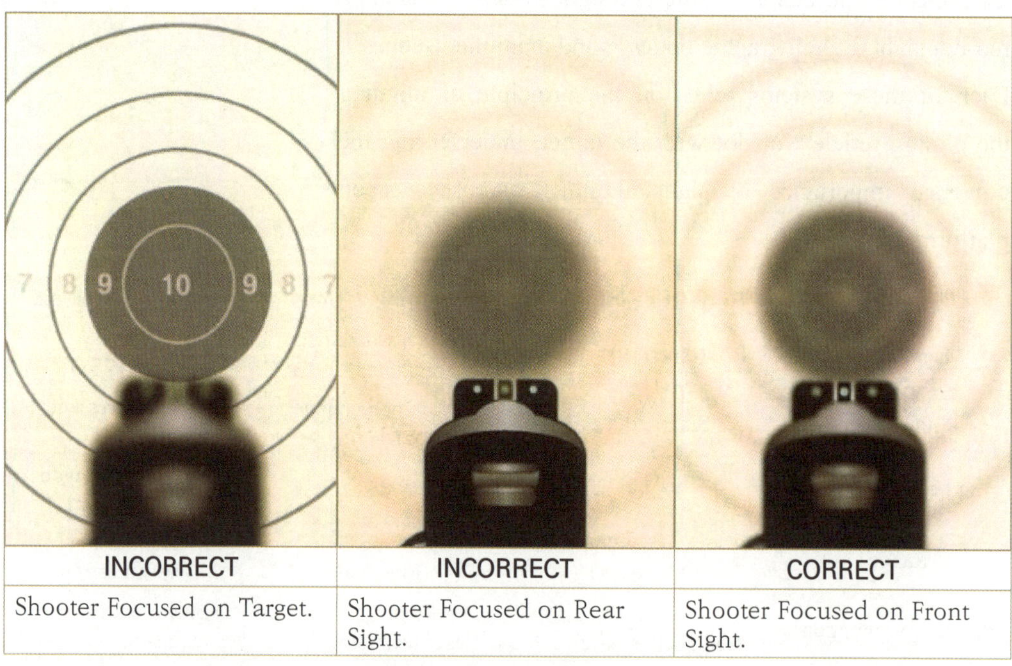

<Exemplar sight pictures>

Creating a proper sight picture requires the shooter to combine precise sight alignment with situational awareness[08] of the target. This involves ensuring that the front sight post is correctly positioned relative to the intended point of aim on the target. The balance of focus is critical, as it allows the shooter to maintain control over their alignment without being distracted by the broader visual field. The ability to achieve a consistent sight picture is particularly important in dynamic scenarios, such as moving targets or engagements with limited visibility. For example, a shooter engaging a running target must still prioritize the front sight post while predicting the target's movement to ensure an accurate shot.

The human eye is naturally limited to focusing on one object at a time. In marksmanship, this principle necessitates prioritizing focus on the front sight post or reticle, rather than the target or rear sight. Maintaining focus on the front sight post ensures that it remains clear and sharp, allowing for precise alignment within the rear aperture. Neglecting this focus can lead to misalignment and subsequent inaccuracy. This practice is particularly critical in engagements where the target may be moving or partially obscured, as the front sight provides a stable point of reference.

C. Maintaining Consistency

Consistency is a cornerstone[09] of accurate shooting. One of the most critical aspects of consistency is achieving a proper stock weld. This term describes the firm and uniform placement of the shooter's cheek on the stock of the weapon. A consistent stock weld aligns the shooter's dominant eye[10] with the sights, eliminating variations that can distort the sight picture. Furthermore, consistent positioning aids in developing muscle memory, allowing the shooter to replicate proper alignment and sight picture under varying conditions.

Consistency extends beyond stock weld to include grip, stance, and trigger control. A firm but relaxed grip ensures the

09 cornerstone : 핵심(주춧돌)

10 dominant eye : 주안(주로 사용하는 눈)

firearm remains steady, while a balanced stance minimizes unnecessary movement. Repeated practice helps shooters build muscle memory, enabling them to instinctively adopt the correct form even in high-stress scenarios.

11 negligible : 무시할 만한
12 visual mass : 가시 영역

For example, a shooter transitioning between targets must maintain consistent positioning to ensure accuracy. Any deviation in stock weld, grip, or alignment can disrupt the sight picture and result in missed shots. Training under simulated stress, such as timed drills or moving targets, reinforces the importance of consistency in these fundamentals.

D. Point of Aim and Desired Point of Impact

The point of aim is the continuation of the line created by sight alignment, directed at a specific location on the target. For stationary targets at ranges under 300 meters, with negligible[11] wind and a confirmed zero at 200 or 300 meters, the point of aim is typically the center of visible mass[12]. This central aim point provides a reliable reference for engaging targets without additional adjustments for hold-off or lead.

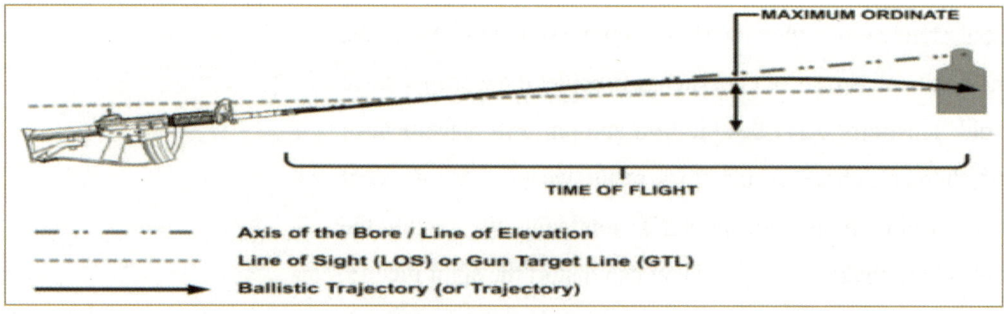

<External ballistic terms>

The desired point of impact, on the other hand, is where the shooter intends the projectile to strike the target. At distances other than the weapon's zero, the point of aim and the desired point of impact may not align due to factors such as bullet drop and environmental conditions. In these cases, the shooter must calculate the necessary hold-off to achieve the desired result. For example, at extended ranges, a shooter may need to aim slightly higher than the intended point of impact to compensate for bullet trajectory. Understanding this relationship allows for more precise adjustments and consistent accuracy.

The relationship between the point of aim and point of impact becomes more complex in varied conditions. Wind, elevation, and target movement all require the shooter to adapt. For instance, when firing in high-wind conditions, the shooter may need to hold off to the left or right of the target, depending on wind direction. Similarly, shooting uphill or downhill affects bullet trajectory, requiring elevation adjustments to maintain accuracy.

Advanced techniques, such as using ballistic calculators or adjusting optics, are often employed to bridge the gap between the point of aim and desired point of impact. These tools provide precise data on bullet drop and environmental factors, enabling shooters to make informed adjustments in real-time.

<A ballistic calculator example>

E. Conclusion

Mastering the key components of proper aiming—sight alignment, sight picture, front sight focus, and consistency—is essential for achieving accuracy in marksmanship. By understanding the dynamics of the point of aim and the desired point of impact, shooters can adapt to various engagement scenarios and environmental conditions. These foundational skills, when practiced and refined, enable a shooter to deliver precise and effective fire, whether on the range or in combat.

Effective marksmanship is not simply a matter of aiming and firing; it is the culmination of disciplined practice, attention to detail, and the application of sound techniques. By combining these principles, shooters can achieve both precision and adaptability, ensuring success in any operational environment.

Part 3. Dialogue

[Dialogue] An instructor and a cadet are at a shooting bench, focusing on mastering sight alignment and sight picture

INSTRUCTOR: (gesturing to the rifle sights) Alright, cadet. Let's focus on sight alignment. What's the first thing you need to do?

CADET: (adjusts the rifle, peering through the sights) Align the front sight post in the center of the rear aperture, with equal light on both sides.

INSTRUCTOR: Good. And where should your focus be?

CADET: (keeps looking through the sights) On the front sight post, not the rear aperture or the target.

INSTRUCTOR: Exactly. The front sight is your guide. A clear post ensures everything else falls into place. Now let's move to the sight picture. What do you combine with sight alignment?

CADET: I align the sights on the target, placing the front sight post on the center of the visible mass.

INSTRUCTOR: Good. And what do you notice about the clarity of the target and sights?

CADET:(adjusting their view) The front sight is sharp, but the target and rear aperture are slightly blurred.

INSTRUCTOR: That's how it should be. Trust the alignment, and the shot will be accurate. [The cadet readies for a shot, exhaling slowly.]

INSTRUCTOR: Maintain your stock weld and steady your aim. Take the shot when you're ready.

CADET:(squeezes the trigger smoothly) <Bang!> How's that?

INSTRUCTOR: (checks through binoculars) On target! Keep practicing this alignment, and it'll become second nature.

Part 4. Quiz

A. Multiple-Choice Questions

01. What should the shooter focus on while aiming?

a. The rear sight aperture
b. The target
c. The front sight post
d. The rifle stock

02. What is the purpose of a proper stock weld?

a. To reduce the rifle's weight
b. To ensure consistent alignment of the dominant eye with the sights
c. To provide extra stability for the rifle barrel
d. To avoid bullet drop during long-range shots

03. What does the sight picture combine?

a. Front sight focus and target motion
b. Sight alignment and placement of the sights on the target
c. Stock weld and sight focus
d. Recoil management and trigger control

04. What adjustment must a shooter make for a long-range target?

a. Focus only on the target
b. Adjust the stock weld position
c. Hold slightly above the point of aim to compensate for bullet drop
d. Move the sights slightly left or right for better clarity

B. True/False Questions

01. Sight alignment involves focusing on the rear sight aperture. (T / F)

02. The front sight post should appear sharp and in focus during proper aiming.(T / F)

03. The sight picture is the combination of sight alignment and the visual placement of the sights on the target.(T / F)

04. The target should be in sharp focus, with the sights slightly blurred.(T / F)

05. A consistent stock weld ensures that the dominant eye remains aligned with the sights. (T / F)

06. The point of aim and the point of impact always align perfectly, regardless of distance. (T / F)

07. The instructor emphasized that accuracy is achieved by prioritizing focus on the target rather than the sights.(T / F)

08. Sight alignment errors can cause significant deviations in the projectile's trajectory.(T / F)

09. The front sight post is centered within the rear aperture to maintain proper sight alignment.(T / F)

10. Proper aiming requires balancing focus between the sights and the environment beyond the target.(T / F)

C. Fill-in-the-Blank Questions

01. A stable __________ is important for accurate shooting because it helps the shooter keep __________ and manage recoil after each shot.

02. The shooter should focus on the __________ to ensure proper alignment, while the target and rear sight appear slightly __________ .

03. __________ is the combination of sight alignment and the visual placement of the sights on the __________ .

04. Maintaining __________ in aiming is achieved by keeping the same __________ on the stock every time.

05. The __________ is the continuation of the line created by __________ and is typically directed at the center of visible mass for stationary targets.

06. If the __________ does not align with the point of aim, the shooter must adjust for __________ or environmental factors.

07. Proper aiming requires aligning the __________ within the __________ to create a straight visual path to the target.

Army stories

조준선 정렬(Sight Alignment)과 표적 정렬(Target Alignment)의 역사적 발전

조준선 정렬과 표적 정렬의 발전은 총기의 역사와 함께 시작되었으며, 이를 구체적으로 설명하기 위해서는 실제 사료와 역사적 사례를 살펴볼 필요가 있다. 초기 화약 무기 시대부터 현대의 정밀 사격술에 이르기까지, 이 두 가지 개념은 군사 기술의 핵심 요소로 자리 잡았다.

초기 화약 무기와 조준선 정렬의 등장[관련 사료: "De Re Militari" (베제티우스, 4세기)]

"De Re Militari"는 전투 기술과 군사 훈련에 관한 고대 로마의 중요한 군사 서적으로, 직접적인 화약 무기에 대한 언급은 없지만, 병사들이 투사 무기의 발사각과 방향을 정밀하게 맞추기 위해 일관된 자세를 유지하는 것이 중요하다고 강조한다. 이는 총기가 등장하기 전부터 표적 정렬의 개념이 존재했음을 보여준다.

15세기 후반에서 16세기 초반, 화약 무기의 발달과 함께 초기 조준선 정렬이 등장했다. **아르퀘버스(Arquebus)**와 같은 초기 총기는 가늠쇠와 가늠자가 없었으며, 사수는 총신의 상단을 따라 목표를 겨냥했다. 사료에 따르면, 1521년 독일의 막스밀리안 1세 훈련 매뉴얼에서는 "총신의 직선을 기준으로 시야를 정렬하라"는 지침이 명시되어 있다. 이는 현대 조준선 정렬의 원형이 되는 개념이었다.

표적 정렬과 정밀 사격의 발전[사료: "English Civil War Drill Manual" (17세기)]

17세기 영국 내전 당시 머스킷병 훈련 매뉴얼에는 "가늠쇠를 가늠자의 중앙에 맞추고, 목표물과 일치시킬 것"이라는 구체적인 조준 지침이 포함되어 있다. 이는 표적 정렬(Target Alignment)이라는 개념이 전장에서 체계적으로 도입되었음을 보여주는 사례다.

표적 정렬은 단순히 조준선 정렬의 연장선이 아니라, 목표물과의 관계를 고려한 정밀 사격의 중요한 단계였다. 당시 사용된 머스킷은 정확도가 떨어졌기 때문에, 병사들은 표적 정렬을 통해 가능한 한 중앙을 겨냥하려고 노력했다. 특히, 일관된 조준을 위해 병사들은 일정한 높이와 간격으로 배열되어 표적 정렬을 단체로 수행하는 방식으로 전술적 우위를 확보했다.

19세기 군사 혁명과 조준선 정렬의 정밀화[사료: "Instruction for Rifle Practice" (1861, 미군)]

19세기에 이르러 조준선 정렬과 표적 정렬은 과학적 근거와 함께 발전하기 시작했다. 특히, 미국 남북전쟁 당시, 저격수들이 사용한 **휘트워스 소총(Whitworth Rifle)**은 1,000야드 이상에서도 높은 정확도를 보였으며, 이는 정밀한 조준선 정렬이 필수적임을 보여주었다.

1861년 발간된 미군의 "Instruction for Rifle Practice"에서는 다음과 같은 조언이 명시되어 있다:

"눈과 가늠쇠, 가늠자를 정확히 일직선으로 정렬하라. 가늠자의 상단은 목표물의 중심과 일치해야 하며, 시야의 일관성이 사격 성공의 열쇠이다."

이 매뉴얼은 조준선 정렬과 표적 정렬이 명확하게 구분된 개념으로 발전했음을 보여준다. 또한, 이러한 훈련은 병사들이 긴장된 전투 상황에서도 정확성을 유지할 수 있도록 돕는 데 중점을 두었다.

현대 전장에서 조준선 정렬과 표적 정렬[사료: TC 3-22.9 "Rifle and Carbine" (미군, 2016년)]

현대 군사 훈련 매뉴얼에서는 조준선 정렬과 표적 정렬을 더욱 정밀하게 다룬다. 미군의 TC 3-22.9는 다음과 같이 설명한다:

"사수는 가늠쇠를 가늠자의 중앙에 맞추고, 동일한 조명과 높이를 유지해야 한다. 조준선이 정렬되면, 이를 목표물과 결합하여 완벽한 표적 정렬을 형성하라."

여기에서 강조된 조준선 정렬과 표적 정렬은 현대의 광학 조준 장치에도 동일하게 적용된다. Red-dot sight와 holographic sight는 기존의 조준 방식을 보완하면서도, 전통적인 정렬 방식의 원리를 유지하고 있다. 이는 기술적 발전이 전통적인 개념을 대체하는 것이 아니라, 이를 보완하고 강화하는 방향으로 발전했음을 보여준다.

06

First Aid and Tactical Combat Casualty Care

By the end of this chapter, you will be able to

- define key first aid and tactical combat casualty care(TCCC) concepts.
- demonstrate the ability to evaluate and respond to a casualty.
- apply appropriate first aid techniques.
- understand vital body systems and their significance in first aid and TCCC.

"The fate of the wounded rests in the hands of the ones who apply the first dressing."

- Dr. Norman McSwain

Part 1. Preview

A. Introduction

When a nonmedical Soldier encounters an unconscious or injured Soldier, he must quickly and accurately evaluate the situation and the casualty to determine if it is safe for him to act as well as what, if any, first aid measures may be needed to prevent further injury or death. To prevent further injury to the casualty, once first aid has begun to be administered there should be no interruptions and those efforts should continue until such time as that Soldier is relieved by medical personnel. Soldiers may also have to depend upon their own first aid knowledge and skills to save themselves (self-aid).

B. Key Doctrinal Terms & Definition

CASEVAC 사상자후송	Casualty evacuation. Nonmedical units use this to refer to the movement of casualties aboard nonmedical vehicles or aircraft without en route medical care.
MEDEVAC 의무후송	Medical evacuation. Medical evacuation is the process of moving any person who is wounded, injured, or ill to and/or between medical treatment facilities while providing en route medical care. Also referred to as MEDEVAC in relation to the 9-line medical evacuation request.
TCCC 전투부상자처치	Tactical combat casualty care(TCCC) is often referred to as TC3. Tactical combat casualty care is prehospital care provided in a tactical-combat setting.
first aid 응급처치	Urgent and immediate lifesaving and other measures which can be performed for casualties (or performed by the casualty himself) by nonmedical personnel when medical personnel are not immediately available.

combat lifesaver 전투구조원	A nonmedical Soldier trained to provide enhanced first aid as a secondary mission.
casualty collection point 전상자수집소	A location that may or may not be staffed, where casualties are assembled for evacuation to a medical treatment facility.
care under fire 교전 중 처치	The care provided by the first TCCC-trained individual in the presence of an immediate threat.
tactical field care 전술적 현장 처치	The care rendered by the first responder or combatant once no longer under effective hostile fire.
tactical evacuation care 전술적 후송 처치	The care that casualties are being transported to an advanced role of care by any means available (air, land, or maritime).

Part 2. Reading

A. First Aid fundamentals

Because medical personnel will not always be readily available, nonmedical Soldiers must rely on themselves and other Soldiers' skills and knowledge of first aid[01] methods to render aid until medical assistance arrives. First aid is given until medical care provided by medically trained personnel such as a combat medic or other health care provider arrives. The individual being provided first aid (by self-aid, buddy aid, or combat lifesaver[02]) is considered a casualty. Once medically trained personnel (combat medic, paramedic, or other health care provider) initiates care, the casualty is then considered to be a patient. A thorough understanding of the fundamentals of first aid can save a life, prevent permanent disability, or reduce long periods of hospitalization by knowing what to do, what not to do, and when to do it.

01 first aid : 응급처치
02 combat lifesaver : 전투구조원
03 respiratory : 호흡기의
04 circulatory : 순환의
05 musculoskeletal : 근골격의
06 compromise : 손상
07 rib cage : 흉곽
08 artery : 동맥
09 capillary : 모세혈관
10 vein : 정맥

Vital Body Systems

Basic understanding of vital body systems is crucial in the understanding and performance of first aid. These systems are the respiratory[03] system, the circulatory[04] system, and the musculoskeletal[05] system. Injury to or compromise[06] of these systems can quickly result in permanent injury or death of the injured. The respiratory system consists of the airway, lungs, and rib cage[07]. The circulatory system provides the vehicle for oxygen acquired through the respiratory process to be transported throughout the body and consists of the heart, blood, and blood vessels(arteries[08], capillaries[09] , and veins[10]

). The musculoskeletal system is composed of bones, joints[11]. muscles, cartilage[12], ligaments[13], and tendons[14].

11 joint : 관절
12 cartilage : 연골
13 ligament : 인대
14 tendon : 힘줄, 건

respiratory system	airway	• Consists of the nose, mouth, throat, voice box, and windpipe. • The canal through which air passes to and fromthe lungs.
	lungs	• two elastic organs made up of thousands of tiny air sacs and covered by an airtight membrane.
	rib cage	• formed by the muscle-connected ribs, which join the spine in back and the breastbone in front.
circulatory system	heart	• the engine that drives the human body by pumping blood.
	blood	• blood is the fuel which powers and sustains the human body by delivering oxygen and essential nutrients (such as fats, sugars, minerals, and vitamins) to the body's tissues. • a mixture of plasma (liquid component), white blood cells, red blood cells, and platelets.
	blood vessels	• the conduit which transports blood pumped by the heart to the body. • consist of arteries, arterioles, capillaries, venules, and veins.
musculoskeletal system	bones	• a firm, rubbery tissue that cushions bones at joints.
	joints	• the areas where two bones are attached for the purpose of permitting body parts to move.
	muscles	• the tissue of the body which primarily functions as a source of power.
	cartilage	• a firm, rubbery tissue that cushions bones at joints.
	ligaments	• a tough band of connective tissue that connects various structures, such as two bones.
	tendons	• a soft tissue by which muscle attaches to bone.

<Vital body systems and its components>

General Principles of First Aid

When a casualty is first encountered it is imperative that the responder quickly and accurately assess what has occurred, determine the nature and extent of injuries and what first aid measures are appropriate and necessary. Transporting or moving a casualty by Soldiers providing first aid must be carefully considered. If there is no danger of greater injury to the casualty by leaving them where they are found, first aid responders should render such aid as is necessary and wait for trained medical personnel to arrive. Evaluation of a casualty is necessary to identify and treat all life-threatening conditions and other serious wounds.

15 Tactical combat casualty care(TCCC) : 전투부상자처치

16 care under fire : 교전 중 처치

17 intervention : 개입, 간섭

B. Tactical Combat Casualty Care

Tactical combat casualty care(TCCC)[15], a set of guidelines for treating battlefield injuries, occurs during a combat mission and is the military counterpart to prehospital emergency medical treatment. TCCC in the military is most commonly provided by enlisted personnel and includes self-aid and buddy aid (first aid), combat lifesaver (enhanced first aid), and enlisted combat medics. TCCC is divided into the three phases : care under fire, tactical field care, and tactical evacuation care.

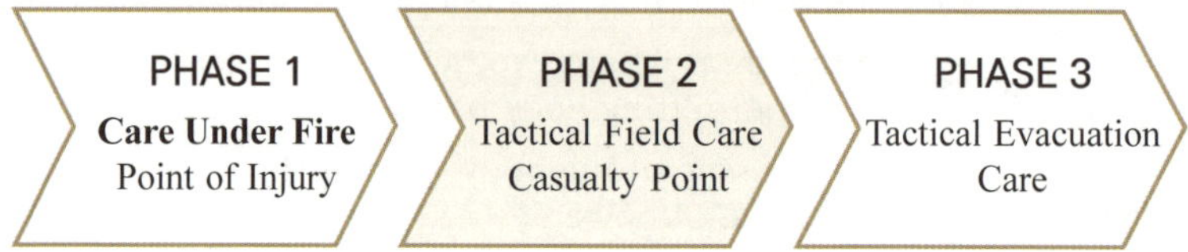

Care Under Fire

In the care under fire(CUF)[16] phase, combat medical personnel and their units are under effective hostile fire and are very limited in the care they can provide. As CUF is conducted while tactically engaged, the most effective way to reduce morbidity and mortality is the precise application of combat fires by all personnel. Continue the tactical mission, gain fire superiority, then treat casualties.In essence, only those lifesaving interventions[17] that must be performed immediately are undertaken during this phase. When you need to perform care under fire, you should :

- Return fire as directed or required before providing first aid.
- Determine if the casualty is alive or dead, and provide care to the living casualty.
- Conduct the following once enemy fire has been suppressed :

(1) In a battle-buddy team, administer lifesaving hemorrhage control[18] by applying a tourniquet[19] from the casualty's Individual First Aid Kit(IFAK)[20] before moving the casualty.

(2) Move the casualty, his weapon, and mission-essential equipment when the tactical situation permits.

18 hemorrhage control : 지혈

19 tourniquet : 지혈대

20 Individual First Aid Kit(IFAK) : 개인 응급처치키트

21 medical treatment facility(MTF) : 의무진료시설

22 casualty collection point : 전상자수집소

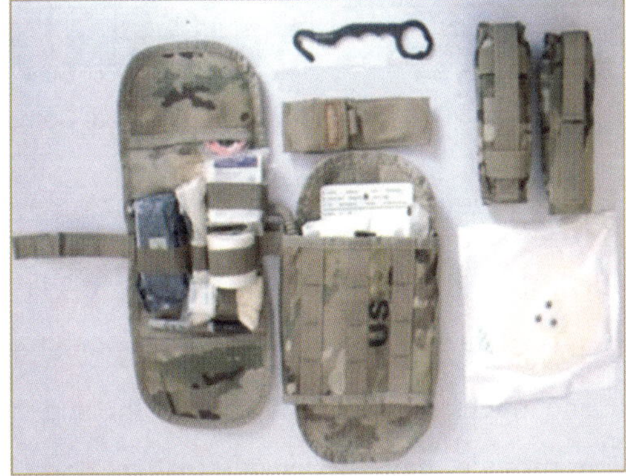

< Combat application tourniquet and IFAK>

All platoon combat lifesaver(CLSs) and the combat medic carry multiple blank versions and use "Tactical Combat Casualty Care Card(TCCC Card)" to document pre-MTF(medical treatment facility)[21] care at the point of injury(POI). Such care relates to both battle and nonbattle injuries. Once completed, TCCC card is visibly attached to the patient when transferred to the casualty collection point(CCP)[22] and/or to MTFs.

TACTICAL COMBAT CASUALTY CARE (TCCC) CARD

BATTLE ROSTER #: ____

EVAC: ☐ Urgent ☐ Priority ☐ Routine

NAME (Last, First): ____ LAST 4: ____

GENDER: ☐ M ☐ F DATE (DD-MMM-YY): ____ TIME: ____

SERVICE: ____ UNIT: ____ ALLERGIES: ____

Mechanism of Injury: (X all that apply)
☐ Artillery ☐ Blunt ☐ Burn ☐ Fall ☐ Grenade ☐ GSW ☐ IED
☐ Landmine ☐ MVC ☐ RPG ☐ Other: ____

Injury: (Mark injuries with an X)

TQ: R Arm TYPE: ____ TIME: ____

TQ: L Arm TYPE: ____ TIME: ____

TQ: R Leg TYPE: ____ TIME: ____

TQ: L Leg TYPE: ____ TIME: ____

S A M P L E

Signs & Symptoms: (Fill in the blank)

Time				
Pulse (Rate & Location)				
Blood Pressure	/	/	/	/
Respiratory Rate				
Pulse Ox % O2 Sat				
AVPU				
Pain Scale (0-10)				

DD Form 1380, JUN 2014 TCCC CARD

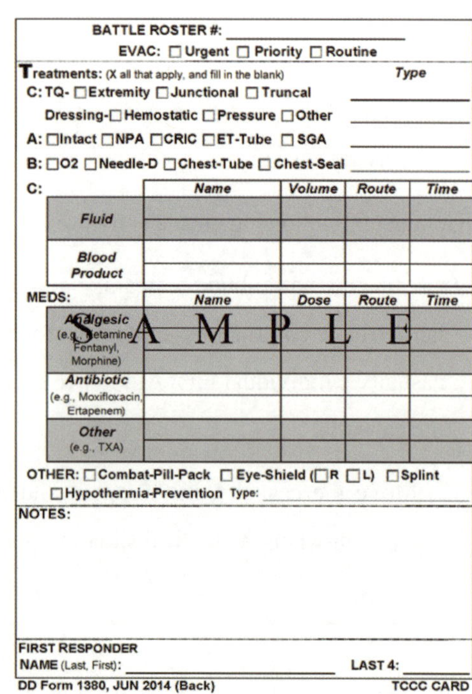

BATTLE ROSTER #: ____

EVAC: ☐ Urgent ☐ Priority ☐ Routine

Treatments: (X all that apply, and fill in the blank) Type

C: TQ- ☐ Extremity ☐ Junctional ☐ Truncal ____

Dressing- ☐ Hemostatic ☐ Pressure ☐ Other ____

A: ☐ Intact ☐ NPA ☐ CRIC ☐ ET-Tube ☐ SGA ____

B: ☐ O2 ☐ Needle-D ☐ Chest-Tube ☐ Chest-Seal ____

C:	Name	Volume	Route	Time
Fluid				
Blood Product				

MEDS:	Name	Dose	Route	Time
Analgesic (e.g., Ketamine, Fentanyl, Morphine)				
Antibiotic (e.g., Moxifloxacin, Ertapenem)				
Other (e.g., TXA)				

S A M P L E

OTHER: ☐ Combat-Pill-Pack ☐ Eye-Shield (☐ R ☐ L) ☐ Splint
☐ Hypothermia-Prevention Type: ____

NOTES:

FIRST RESPONDER
NAME (Last, First): ____ LAST 4: ____

DD Form 1380, JUN 2014 (Back) TCCC CARD

<DD form 1380, Tactical Combat Casualty Care [TCCC] Card>

Tactical Field Care

Tactical field care[23] is the backbone of TCCC. During the tactical field care phase, medical personnel and their patients are no longer under effective hostile fire and medical personnel can provide more extensive patient care. In this phase, interventions directed at other life-threatening conditions, as well as resuscitation[24] and other measures to increase the comfort of the patient may be performed. During tactical field care, personnel must be prepared to transition back to care under fire, or to prepare the casualty for tactical evacuation, as the tactical situation dictates. TCCC teaches first responders to treat casualties in the proper order, treating the most critical situations first. This is done using the acronyms MARCH and PAWS.

23 tactical field care : 전술적 현장 처치

24 resuscitation : 소생

MARCH Acronym
(massive hemorrhage, airway, respirations, circulation, head injury/hypothermia)

- Massive hemorrhage[25]

The number one potentially survivable cause of death at the POI is hemorrhage from a compressible wound or any life-threatening extremity bleed. The hasty application of a tourniquet[26] is the recommended management for all life-threatening extremity hemorrhages during care under fire. It is initially placed over clothing, high and tight. The deliberate application of a tourniquet is addressed when behind cover and during tactical field care to ensure proper hemorrhage control. The tourniquet is placed under clothing 2 to 3 inches above the wound. The application time is written on the tourniquet.

- Airway[27]

A second survivable cause of death at the POI is a non-patent(closed) airway. A conscious and speaking casualty has a patent open airway. An unconscious casualty who is breathing can benefit from the nasopharyngeal airway(NPA)[28].

- Respirations[29]

The third potentially survivable cause of death on the battlefield is the development of a tension pneumothorax(PTX)[30]. Air trapped in the chest cavity begins to displace functional lung tissue and places pressure on the heart, resulting in cardiac arrest. Seal open chest wounds with a vented chest seal[31] , decompress a suspected PTX, and support ventilation/oxygenation, as required.

25 hemorrhage : 출혈
26 tourniquet : 지혈대
27 airway : 기도
28 nasopharyngeal airway : 코인두기도기(비강인두관)
29 respiration : 호흡
30 tension pneumothorax : 긴장성 기흉
31 vented chest seal : 통기성 흉부 씰

- Circulation[32]

Control of bleeding takes precedence over infusing fluids. Only individuals in shock or those who need intravenous(IV)[33] medications need to have IV access established.

- Head injury / hypothermia[34]

Non-medical personnel utilize the alert, verbal, pain, unresponsive(AVPU)[35] scale to identify traumatic brain injury (TBI). Hypothermia is a survivable cause of further injury and is defined as a whole body temperature below 35°C. Hypothermia can occur secondary to blood loss, regardless of the ambient temperature. The Hypothermia Prevention and Management Kit (HPMK) is recommended for all casualties.

PAWS Acronym(pain, antibiotics, wounds, splinting)

- Pain

Management of a casualty's pain helps reduce stress on the mind and body. Pain management reduces patient movement, improves compliance and cooperation, and allows for better casualty transport and outcomes.

- Antibiotics[36]

Antibiotics are used to treat multidrug-resistant bacteria. All battlefield wounds are considered contaminated. All wounded casualties with penetrating injuries should receive antibiotics.

- Wounds

Assessing and treating casualties with additional wounds improves morbidity and mortality. Minor scalp lacerations

32 circulation : 순환(혈액)
33 intravenous(IV) : 정맥주사
34 hypothermia : 저체온증
35 AVPU : 부상자의 의식수준을 평가하는 방법
36 antibiotics : 항생제

can be the cause of excessive hemorrhage. First responders must address burns, open fractures, facial trauma, amputation dressings, and security of tourniquets. Reassessment of wounds and interventions prior to movement is critical.

- Splinting[37]

Medical personnel should address pelvic[38] disruptions and eye injuries. Splinting of fractures can result in significant pain relief and minimize bleeding.

Tactical Evacuation Care

In the tactical evacuation care[39] phase, casualties are transported from the battlefield to MTFs. MTFs provide medical treatment and include the Role 1 facility (battalion aid station), Role 2 facility (brigade support medical company of the brigade support battalion), dispensaries, clinics, and hospital. Evacuation can be by either Medical evacuation(MEDEVAC)[40] or Casualty evacuation(CASEVAC)[41].

CASEVAC is the movement of casualties aboard nonmedical vehicles or aircraft without en route medical care. CASEVAC encompasses a wide spectrum of potential capability—depending on the mix of transport platform, medical equipment, and medical providers allocated to the mission. At the upper end of the spectrum, nondedicated platforms can be outfitted with the requisite medical equipment and MEDEVAC assets. Effective CASEVAC complements MEDEVAC by providing additional evacuation capacity when number of casualties (workload) or reaction time exceeds the capabilities of MEDEVAC assets. CASEVAC requires detailed assessment and planning to achieve an effective integration of MEDEVAC and CASEVAC capabilities.

37 splint : 부목
38 pelvic : 골반
39 tactical evacuation care : 전술적 후송 처치
40 Medical evacuation(MEDEVAC) : 의무후송
41 Casualty evacuation(CASEVAC) : 사상자후송

MEDEVAC is the timely and effective movement of the wounded, injured, or ill to and between medical treatment facilities on dedicated and properly marked medical platforms with en route care provided by medical personnel. MEDEVAC is the key factor to ensuring the continuity of care provided to Soldiers by providing en route medical care during evacuation and facilitating the transfer of patients between MTFs to receive the appropriate specialty care. This ensures that scarce medical resources can be rapidly transported to areas of critical need on the battlefield.

Part 3. Dialogue

[Dialogue 1] A platoon sergeant and his soldier are discussing Tactical Combat Casualty Care (TCCC)

PV KIM: Sergeant, as far as I know, evacuation can be by either Medical evacuation(MEDEVAC) or Casualty evacuation(CASEVAC). what's the difference between CASEVAC and MEDEVAC? I always mix them up.

SGT LEE: Good question, Kim. CASEVAC is casualty evacuation. It's when we move injured personnel using nonmedical vehicles or aircraft, and there's no en route medical care.

PV KIM: Got it. And MEDEVAC?

SGT LEE: MEDEVAC stands for medical evacuation. It involves dedicated medical platforms with proper markings, and it includes en route care by trained medical personnel.

PV KIM: So, MEDEVAC is more specialized and provides medical treatment during the transport?

SGT LEE: Exactly. MEDEVAC ensures continuity of care from the point of injury to medical treatment facilities. It's critical for saving lives on the battlefield. But, keep that in mind that effective CASEVAC complements MEDEVAC by providing additional evacuation capacity when number of casualties or reaction time exceeds the capabilities of MEDEVAC assets.

PV KIM: That means detailed assessment and planning on CASEVAC must be considered to achieve an effective integration of MEDEVAC and CASEVAC capabilities, right?

SGT LEE: Exactly. I hope my answers clear things up.

PV KIM: Thanks, Sergeant.

[Dialogue 2] A cadet asks MARCH Acronym in Tactical Field Care during his First Aid and TCCC class.

CADET PARK: Sir, could you explain the MARCH acronym in TCCC? I'm still not confident about applying it.

LT CHOI: Of course, Park. MARCH stands for Massive hemorrhage, Airway, Respirations, Circulation, and Head injury / hypothermia. These are the priorities in tactical field care.

CADET PARK: So, if I find a casualty with heavy bleeding and difficulty breathing, which comes first?

LT CHOI: You manage massive hemorrhage first. Apply a tourniquet high and tight to stop the bleeding should be done first.

CADET PARK: What about the airway?

LT CHOI: Once the bleeding is controlled, you ensure the airway is open. If the casualty is unconscious but breathing, you can use a nasopharyngeal airway.

CADET PARK: And for respiration issues like tension pneumothorax?

LT CHOI: You'll need to seal open chest wounds with a vented chest seal and decompress if necessary. Always follow the sequence—address the most critical issues first.

CADET PARK: Understood, sir. I'll practice this sequence.

LT CHOI: Good. You should know all items and its usage in your IFAK as a combat lifesaver. Confidence comes with practice, Park. You're doing well.

Part 4. Quiz

A. Multiple-Choice Questions

01. What is the first priority in the MARCH acronym for TCCC?

a. Airway
b. Circulation
c. Massive hemorrhage
d. Head injury

02. Which of the following is NOT part of the circulatory system?

a. Blood vessels
c. Lungs
b. Heart
d. Blood

03. What does AVPU stand for in evaluating a casualty?

a. Alert, Visual, Pain, Unconscious

b. Alert, Verbal, Pain, Unresponsive

c. Airway, Vision, Pain, Urgency

d. Awareness, Voice, Pain, Unresponsive

04. What is a key feature of tactical evacuation?

a. Only medical personnel are allowed to transport casualties.

b. It focuses on evacuation using only medical vehicles.

c.It includes CASEVAC and MEDEVAC.

d. It ensures only minor injuries are evacuated.

05. What should be the immediate action under "Care Under Fire"?

a. Return fire and suppress the enemy.

b. Conduct a thorough medical assessment.

c. Apply a splint.

d. Move the casualty to the nearest MTF.

B. True/False Questions

01. Nonmedical Soldiers are not required to administer first aid if they believe medical personnel are nearby. (T / F)

02. CASEVAC involves en route medical care provided to casualties during their evacuation. (T / F)

03. Tactical combat casualty care (TCCC) can include self-aid and buddy aid. (T / F)

04. The human body has 206 bones, which form the musculoskeletal system. (T / F)

05. A casualty with an open chest wound should be treated using a vented chest seal.(T / F)

C. Fill-in-the-Blank Questions

01. The ______________ system consists of the heart, blood, and blood vessels.

02. A ______________ is used to control life-threatening extremity bleeding during care under fire.

03. The primary survivable cause of death at the point of injury (POI) is ______________ from a compressible wound.

04. ______________ is the military counterpart to prehospital emergency medical treatment.

05. The ______________ phase of TCCC focuses on patient care when medical personnel and patients are no longer under effective hostile fire.

D. Matching Questions

01. Match the body systems to their primary components:

1. Heart, blood vessels, blood	a. Respiratory system
2. Bones, joints, muscles	b. Circulatory system
3. Airway, lungs, rib cage	c. Musculoskeletal system

02. Match the phases of TCCC to their characteristics:

4. Lifesaving measures under hostile fire	d. Care under fire
5. Extended care when no longer under fire	e. Tactical evacuation care
6. Transporting casualties to MTFs	f. Tactical field care

E. Scenario-based Question

01. A Soldier has a penetrating chest wound. What is the correct first aid measure to take?

a. Apply a tourniquet.
b. Seal the wound with a vented chest seal.
c. Give antibiotics immediately.
d. Start CPR.

02. You encounter a casualty with uncontrolled bleeding and a closed airway during tactical field care. What should you prioritize according to the MARCH acronym?

a. Treat the airway first, then manage hemorrhage.
b. Wait for medical personnel.
c. Treat both simultaneously.
d. Address hemorrhage first, then treat the airway.

03. During evacuation, which type of transport provides en route medical care?

a. MEDEVAC
b. CASEVAC
c. TACEVAC
d. Combat lifesaver

Army stories

How to apply a tourniquet (지혈대 사용법)

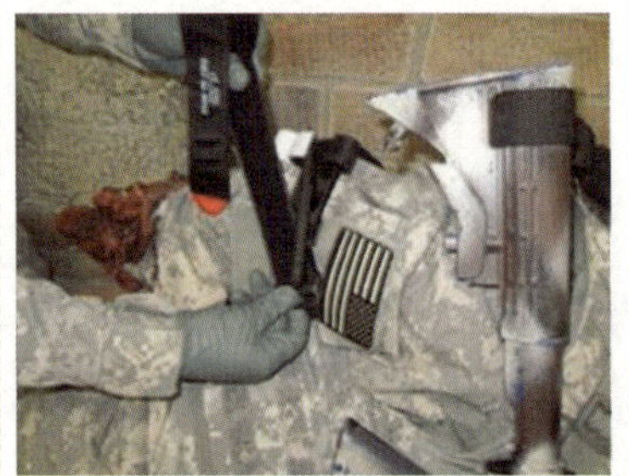	**Step 1.** 교전 중 처치(Care Under Fire) 단계에서는 지혈대를 부상당한 팔다리 부위의 가능한 한 높은 곳(심장에 가까운곳)에, 그리고 전투복 위에 적용한다. 이는 전투 중 추가적인 사상자 발생에 대비한 임시 지혈대(Hasty Tourniquet) 조치이다.
	**Step 2.** 벨크로 밴드(조임끈)을 버클 안에서 밖으로 통과시키고 밴드를 팽팽히 당겨 지혈부위가 압박이 되도록 한후, 벨크로를 부착하여 고정한다. 이때, 밴드와 부상 부위 사이에는 세 손가락 이상 들어가지 않도록 한다.
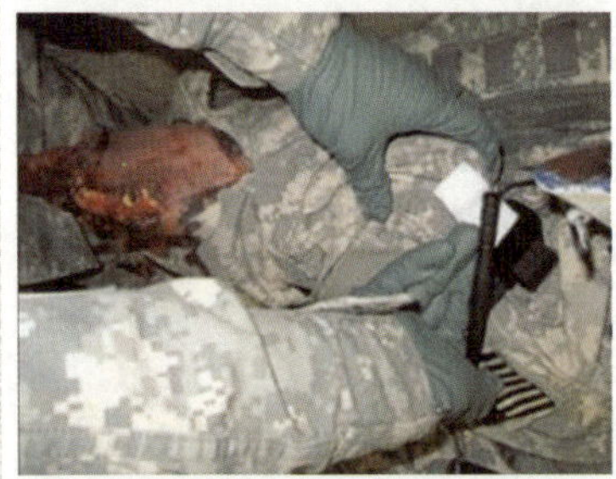	**Step 3.** 조임막대를 돌려 출혈이 완전히 멈출 때 까지 돌려준 후, 조임막대 고정대에 끼워 고정한다. (이 단계까지 1분 이내에 처치한다.)
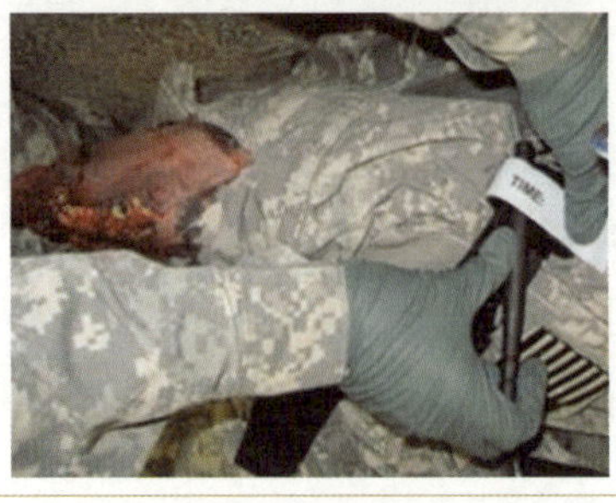	**Step 4.** 이후 조정대 덮개를 부착하여 조임막대를 고정한다. 교전 중 처치 단계에서는 시간을 기록하지 않으며, 전술적 현장 처치 단계에서 정밀 지혈대 처치시 24시간 표기법으로 지혈대 처치 시간을 기록한다.

* 전술적 현장 처치(Tactical Field Care) 단계에서, 부상 부위로부터 2인치 위 피부 위로 옮겨 고정하는 정밀 지혈대(Deliberate Tourniquet) 처치로 전환한다.

Individual First Aid Kit (IFAK) 개인응급처치 키트

미군은 모든 군인에게 개인응급처치키트(IFAK)를 지급하고 있으며, 즉각적인 자가 응급처치와 전우 응급처치를 위해 IFAK의 모든 구성품과 사용법에 대한 교육을 하고 있다. IFAK의 구성품은 파우치(Utility Pouch), 지혈대(Combat Application Tourniquet) 2개, 탄력압박붕대(Elastic Bandage Kit), 거즈(Bandage Gauze), 외과용 테이프(Surgical Adhesive Tape), 코인두기도기(Nasopharyngeal Airway), 의료용 장갑 4켤레(Patient Exam Gloves), 드레싱 거즈(Dressing, Combat Gauze), 스트랩 커터(Strap Cutter, Rescue), 흉부 씰 패치(Bolin Chest Seal), 안구 씰 패치(Eye Shield, Fox), 마커(Marker, Tube Type)이다.

2013년 이래로 미군은 US Army Improved First Arid Kit (IFAK Gen II)를 사용하고 있다.

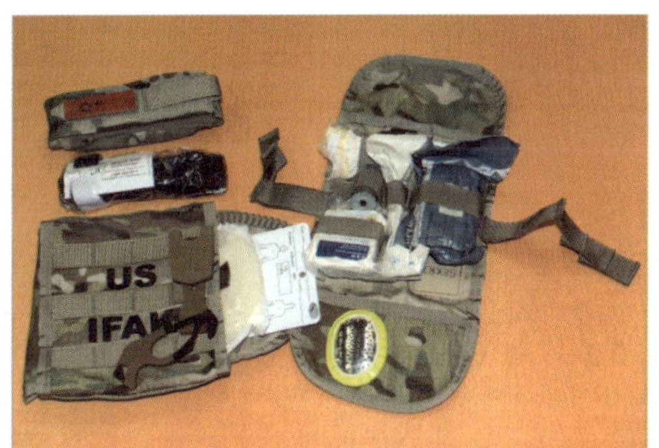

07

Soldier Combat Skills Ⅰ

Cover, Concealment, and Camouflage

By the end of this chapter, you will be able to

- explain the Characteristics of the Battlefield and Survivability.
- define the Cover, Concealment, and Camouflage.
- distinguish the differences among them.

"The most important thing in war is not victory, but survival."

- Nelson Mandela -

Part 1. Preview

A. Introduction

If the enemy can see you and you are within range of his weapon system[01], he can engage and possibly kill you. So, you must be concealed from enemy observation and have cover from enemy fire. When the terrain does not provide natural cover[02] and concealment[03], you must prepare your cover and use natural and man-made materials[04] to camouflage[05] yourself, your equipment, and your position. This chapter provides guidance on the preparation and use of cover, concealment, and camouflage.

01 weapon system : 무기체계
02 cover : 엄폐
03 concealment : 은폐
04 natural and man-made materials : 자연 및 인공 재료
05 camouflage : 위장

B. Key Doctrinal Terms & Definition

Term	Definition
cover 엄폐	any physical barrier or protection that can stop or deflect enemy fire or shrapnel, providing safety during combat.
concealment 은폐	the use of natural or artificial objects to hide personnel or equipment from enemy observation. Unlike cover, concealment does not provide protection against fire.
camouflage 위장	the disguising of military personnel, equipment, and installations by painting or covering them to make them blend in with their surroundings.
trench 교통호, 참호, 도랑	a long, narrow excavation in the ground, typically used in military or construction
camouflage nets 위장막	a textile covering used to conceal or disguise people, objects, or structures from detection.
biological agents 생물학적 작용제	a microorganism or toxin derived from a biological source that can be used as a weapon in biological warfare or bioterrorism.

chemical agents 화학적 작용제	a chemical compound or element that can have harmful effects on living organisms.
SOP 예규	Standard Operating Procedure. A set of instructions applicable to those features of operations that lend themselves to a definite or standardized procedure without loss of effectiveness.

Part 2. Reading

A. Characteristics of the Battlefield and Survivability

Modern combat is chaotic[06], intense[07], and shockingly destructive. In your first battle, you will experience the confusing and often terrifying sights, sounds, smells, and dangers of the battlefield—but you must learn to survive and win despite them.

(1) You could face a fierce[08] and relentless[09] enemy.

(2) You could be surrounded by destruction and death.

(3) Your leaders and fellow soldiers may shout urgent commands and warnings.

(4) Rounds[10] might impact near you.

(5) The air could be filled with the smell of explosives and propellant[11].

(6) You might hear the screams of a wounded comrade[12].

However, even in all this confusion and fear, remember that you are not alone. You are part of a well-trained team, backed by the most powerful combined arms force, and the most modern technology in the world.

06 chaotic : 혼란스러운, 무질서한
07 intense : 격렬한, 치열한
08 fierce : 맹렬한, 몹시 사나운
09 relentless : 가차없는, 잔인한
10 rounds : 총알, 탄환
11 propellant : (로켓 등의) 추진체
12 comrade : 전우, 동료
13 fragments of exploding rounds : 폭발하는 탄약의 파편들
14 frame : 화염
15 biological and chemical agents : 생물학적 및 화학적 작용제

B. Cover

Cover, made of natural or man-made materials, gives protection from bullets, fragments of exploding rounds[13], flame[14], nuclear effects, biological and chemical agents[15], and enemy observation. Natural cover includes logs, trees, stumps[16], rocks, and ravines[17]; whereas, man-made cover includes fighting positions, trenches[18], walls, rubble[19], and craters[20]. To get protection from enemy fire in the offense or when moving, use routes that put cover between you and the enemy. For example, use ravines, gullies[21], hills, wooded areas, walls, and any other cover that will keep the enemy from seeing and firing at you. Avoid open areas. Never skyline yourself on a hilltop[22] or ridge[23]. Any cover (even the smallest depression[24] or fold[25] in the ground) can help protect you from direct and indirect[26] enemy fire.

16 stump : (나무의) 그루터기
17 ravine : 산골짜기, 협곡
18 trench : 참호, 도랑
19 rubble : 돌무더기
20 crater : 분화구, 큰 구멍
21 gully : 도랑, 배수로
22 hilltop : 언덕 꼭대기
23 ridge : 능선
24 depression : 함몰, 움푹 파인 곳
25 fold : 주름, 우묵한 곳
26 direct and indirect : 직 · 간접

< Figure 7-1. Natural cover>

< Figure 7-2. Man-made cover>

Three different types of cover (overhead, frontal, and flank[27]/rear cover) are used to make fighting positions. In addition, positions can be connected by tunnels and trenches. These allow Soldiers to move between positions for engagements or resupply[28], while remaining protected.

27 flank : 측면, 옆구리
28 resupply : 재보급

OVERHEAD COVER (OHC)

Your completed position should have overhead cover, which enhances survivability by protecting you from indirect fire and fragmentation.

FRONTAL COVER

Your position needs frontal cover to protect you from small arms fire to the front. Frontal cover allows you to fire to the oblique[29], as well as to hide your muzzle flash[30].

29 oblique : 사선, 45°
30 muzzle flash : 총구 섬광

FLANK AND REAR COVER

When used with frontal and overhead cover, flank and rear cover protects you from direct enemy and friendly fire. Natural frontal cover such as rocks, trees, logs, and rubble is best, because it is hard for the enemy to detect. When natural cover is unavailable, use the dirt you remove to construct the fighting position. You can improve the effectiveness of dirt as a cover by putting it in sandbags. Fill them only three-quarters full.

C. Concealment

Concealment is anything that hides you from enemy observation (Figure 7-3). Concealment does not protect you from enemy fire. Do not think that you are protected from the enemy's fire just, because you are concealed. Concealment, like cover, can also be natural or Soldier made.

< Figure 7-3. Concealment>

Natural concealment includes bushes[31], grass, and shadows. If possible, natural concealment should not be disturbed. Man-made concealment includes Army combat uniforms (ACUs)[32], camouflage nets[33], face paint, and natural materials that have been moved from their original location. Man-made concealment must blend into natural concealment provided by the terrain.

31 bush : 관목, 덤불
32 Army combat uniforms (ACUs) : 디지털 전투복
33 camouflage nets : 위장막

NATURAL, UNDISTURBED MATERIALS

Natural, undisturbed concealment is better than man-made concealment. While digging your position, try not to disturb the natural concealment around it. Put the unused dirt from the hole behind the position and camouflage it. Camouflage material that does not have to be replaced (rocks, logs, live bushes, and grass) is best. Avoid using so much camouflage that your position looks different from its surroundings. Natural, undisturbed concealment materials--

- Are already prepared.
- Seldom attract enemy attention.
- Need no replacement.

MAN-MADE CONCEALMENT

Your position must be concealed from enemy aircraft as well as from ground troops. If the position is under a bush or tree, or in a building, it is less visible from above. Spread leaves, straw, or grass on the floor of the hole to keep freshly dug earth from contrasting with the ground around it. Man-made concealment must blend with its surroundings so that it cannot be detected, and must be replaced if it changes color or dries out.

D. Camouflage

Camouflage is anything you use to keep yourself, your equipment, and your position from being identified. Both natural and man-made material can be used for camouflage. Change and improve your camouflage often. The time between changes and improvements depends on the weather and on the material used. Natural camouflage will often die, fade, or otherwise lose its effectiveness. Likewise, man-made camouflage may wear off or fade and, as a result, Soldiers, their equipment, and their positions may stand out[34] from their surroundings[35].

34 stand out : 눈에 띄다
35 surrounding : 주변 환경

To make it difficult for the enemy to spot them, Soldiers should remember the following when using or wearing camouflage.

(1) MOVEMENT : Movement and activity draw attention. When you give arm-and-hand signals or walk about your position, your movement can be seen by the naked eye at long ranges. In the defense, stay low. Move only when necessary. In the offense, move only on covered and concealed routes.

(2) POSITIONS : Avoid putting anything where the enemy expects to find it. Build positions on the side of a hill, away from road junctions or lone buildings, and in covered and concealed places. Avoid open areas.

(3) OUTLINES AND SHADOWS : Outlines and shadows can reveal your position or equipment to an air or ground observer. Break up outlines and shadows with camouflage. When moving, try to stay in the shadows.

(4) SHINE : A shine will naturally attract the enemy's attention. In the dark, a burning cigarette or flashlight will give you away. In daylight, reflected light from any polished surface

such as shiny mess gear[36], a worn helmet, a windshield, a watch crystal and band, or exposed skin will do it.

(5) SHAPE : Certain shapes, such as a helmet or human being, are easily recognizable. Camouflage, conceal, and break up familiar shapes to make them blend in with their surroundings, but avoid overdoing it.

(6) COLORS : If your skin, uniform, or equipment colors stand out against the background, the enemy can obviously detect you more easily than he could otherwise. For example, ACUs stand out against a backdrop[37] of snow-covered terrain. Once again, camouflage yourself and your equipment to blend with the surroundings.

(7) DISPERSION : This means spreading Soldiers, vehicles, and equipment over a wide area. The enemy can detect a bunch of Soldiers more easily than they can detect a lone Soldier. Spread out. Unit SOP[38] or unit leaders vary distances between you and your fellow Soldiers depending on the terrain, degree of visibility, and enemy situation.

(8) PREPARATION : Before camouflaging, study the terrain and vegetation[39] of the area in which you are operating. Next, pick and use the camouflage material that best blends with the area (Figures 7-5). When moving from one area to another, change camouflage as needed to blend with the surroundings. Take grass, leaves, brush, and other material from your location and apply it to your uniform and equipment, and put face paint on your skin.

36 mess gear : 군용 식기
37 backdrop : 배경
38 SOP (Standard Operating Procedure) : 예규
39 vegetation : 초목, 식물

<Figure 7-4. Soldier in arctic camouflage>

<Figure 7-5. Camouflaged Soldiers.>

Part 3. Dialogue

[Dialogue] The instructor and cadet are having a conversation during class.

INSTRUCTOR: What is the difference between cover and concealment?

CADET: I think the main difference between cover and concealment is the type of protection they provide. Cover provides physical protection from enemy attacks, such as bullets, explosions, or shrapnel while concealment provides visual protection, hiding someone from the enemy's sight but offering no physical defense.

INSTRUCTOR: You know exactly. Then, can you give an example of cover and concealment?

CADET: Hmm. Hiding behind a concrete wall to avoid bullets is cover, and hiding in tall grass to remain unseen is concealment.

INSTRUCTOR: You understand the point very well. And you also learned about camouflage, didn't you?

CADET: Yes, sir. You explained it last time. Can I ask you a question?

INSTRUCTOR: Of course. Go ahead.

CADET: I would like to know the difference between camouflage and concealment.

INSTRUCTOR: The difference between concealment and camouflage lies in their methods and purposes. The purpose of camouflage is to make a person, object, or vehicle harder to detect by blending it into the environment. This is achieved by using colors, patterns, or materials that closely resemble the surroundings. However, concealment is to make someone or something physically undetectable from an enemy's view by using positioning or obstructions. Then I'll give you a question. Distinguish between camouflage and concealment in the following sentence. "A soldier hides behind a bush while wearing a camouflaged uniform to blend into the vegetation."

CADET: Concealment is hiding behind a bush, and camouflage is wearing a camouflaged uniform to blend into the vegetation. Am I correct?

INSTRUCTOR: Excellent!! You understand it very well.

Part 4. Quiz

A. Multiple-Choice Questions

01. Which of the following is a method of protecting against enemy observation and fire?

a. Cover

b. Army combat uniforms

c. Camouflage

d. Concealment

02. What is the purpose of cover?

a. To protect against physical attacks like bullets, shrapnel, or explosions.

b. To hide your presence or location from the enemy's view.

c. To prevent the enemy from detecting or targeting the individual.

d. To disrupt recognition by blending into the environment.

03. What is not an example of cover?

a. A soldier takes position behind a concrete wall to block incoming bullets.

b. A boulder in a field protects against both gunfire and shrapnel.

c. Covering equipment or personnel with netting that blends with the environment.

d. A tank or military vehicle provides physical protection from enemy attacks.

04. What is not an example of camouflage?

a. Painting tanks with desert colors in a sandy terrain.

b. Wearing forest-patterned uniforms in a wooded area.

c. Using netting or materials to hide equipment or bunkers.

d. Using fog or smoke grenades to obscure movement.

B. True/False Questions

01. You must be covered from enemy observation and have conceal from enemy fire. (T / F)

02. Cover can be only made of natural materials. (T / F)

03. Concealment protect you from enemy fire. (T / F)

04. Man-made cover includes fighting positions, trenches, walls, rubble, and craters. (T / F)

05. Concealment, like cover, can also be natural or Soldier made. (T / F)

06. Do not change and improve your camouflage often. (T / F)

07. Man-made concealment must blend into natural concealment provided by the terrain. (T / F)

08. The time between changes and improvements depends on the weather and on the material used. (T / F)

09. In the dark, you don't need to worry about a shine(T / F)

10. Concealment is anything you use to keep yourself, your equipment, and your position from being identified. (T / F)

C. Fill-in-the-Blank Questions

01. ________ gives protection from bullets, fragments of exploding rounds, flame, nuclear effects, biological and chemical agents, and enemy observation.

02. To get protection from enemy fire in the offense or when moving, use routes that put cover between _______ and _______.

03. Any cover (even the smallest depression or fold in the ground) can help protect you from ______ and ______ enemy fire.

04. __________ is anything that hides you from enemy observation.

05. Outlines and __________ can reveal your position or equipment to an air or ground observer.

06. __________ is anything you use to keep yourself, your equipment, and your position from being identified.

07. Both __________ and __________ material can be used for camouflage.

Army stories

은폐 및 위장의 대명사 '전투복'의 발전

군의 은폐 및 위장 기술은 전장의 환경 변화와 기술적 진보에 따라 지속적으로 발전해 왔다. 이러한 기술은 병력과 장비의 생존성을 높이고 전술적 우위를 확보하는 데 핵심적인 역할을 한다.

그중 전투복은 군복의 일종으로, 전투용으로 만들어진 의복을 의미하는데 생존을 위해 위장 및 은폐 기술이 다양하게 적용되어 있다. 전투복의 역사는 국가별로 다양하게 나타난다. 우리나라는 조선시대부터 광복 전까지 당시에 무기체계 및 의복기술에 맞게 여러 형태의 군복을 착용해 왔다.

기간	창군 (1946.1) ~1965	1965 ~1976	1977 ~1990	1991~2010	2011~2018	2019~
군복	복제규정 없이 혼착 군복 (일본군 군복, 미군원품, 국산 작업복 등)	국방색 전투복		얼룩무늬 전투복 (IR 위장 적용)	디지털무늬 전투복 (항균, 흡한속건 등 기능성 적용 및 패턴/ 치수체계 변경)	디지털무늬 전투복 (소재 변경 및 야간위장 범위 확대)
소재	면 100%	면 100%	폴리에스터 50%, 면 50%	폴리에스터 65%, 면 35%	폴리에스터 68%, 면 32%	폴리에스터 73%, 레이온 27%
				하계: 폴리에스터 65%, 레이온35% (1995~2010)	하계: 폴리에스터 65%, 레이온 35% (2013~2018)	하계: 폴리에스터 70%, 레이온 30% (2019~)

하지만 실질적으로 복식규정에 따라 군복이라고 지칭하기 시작한 것은 1897년(고종)에 서양식 군복을 최초 도입하면서부터였다. 이후 60년대 중반까지는 미군과 일본군의 군복을 모방해 입었고, 명칭 또한 작업복(作業服)이라고 불렀다. 1967년이 되어서야 전투복이란 명칭을 사용하면서, 이후 색상, 소재 등에서 많은 발전을 거듭하게 된다.

특히, 2010년대에 들어서면서 한반도 지형과 기후 특성을 고려하여 흙•침엽수•수풀•나무줄기•목탄의 5가지 색상과 화강암 무늬를 바탕으로 위장 효과가 강화된 디지털무늬 전투복이 도입되어 지금까지 착용하고 있다. 신형 디지털무늬 전투복은 향균, 흡한속건 등 기능성이 적용되었으며, 상의 앞 여밈 방식이 기존 단추에서 지퍼 및 접착포로 바뀌었다.

세계 각국에서는 전투복에 첨단 과학기술을 접목한 차세대 전투복 개발에 박차를 가하고 있다. 신소재 원단과 나노 머신 및 착용형 기기를 바탕으로 한 스텔스 기술 전투복이 바로 그 예이다. 특히 스텔스 망토는 현재 세계 각국에서 활발한 연구가 이루어지고 있으며 늦어도 2030년대에는 실용화가 가능할 것으로 예상되고 있다.

<스텔스 망토>

스텔스 기술이 전투복에 적용되면 광학위장(Optical Camouflage) 또는 능동위장(Active Camouflage)이라 불리는 시각은폐기술을 사용하여 자연스럽게 주변 환경과 동화되거나 시각적으로 존재 자체를 완전히 숨길 수 있게 된다.

<스텔스 기술 전투복>

영화 '아이언맨'에서는 주인공이 '아이언맨 슈트'라고 불리는 웨어러블 로봇(Wearable Robot)방식의 강화복이 등장한다. 웨어러블 로봇은 위상보다는 기능 및 무장 강화에 중점을 둔 기술이다. 일부 사람들은 웨어러블 로봇기술이 발달함에 따라 기존 전투복 개선의 필요성이 다소 낮아질 것이라 주장한다. 하지만 웨어러블 로봇기술 발달과 더불어 전투복 소재의 발전도 혁신적으로 이루어지고 있어 가벼우면서 방탄기능이 적용된 전투복 소재도 개발되었고, 3D 스캔 기능을 이용하여 맞춤형 군복도 생산되고 있다. 전투복은 웨어러블 로봇기술과 더불어 지속적으로 발전할 것이며, 머지않아 우리 군도 스텔스 전투복이나 특수 소재가 적용된 전투복을 착용하고 임무를 수행할 것이라 기대해 본다.

08

Soldier Combat Skills Ⅱ

Movement

By the end of this chapter, you will be able to

- explain the Individual Movement Techniques.
- define the Low crawl, High crawl, and Rush.
- understand the Movement Formations.

"The most important thing in war is movement. Only by continuing to advance without stopping can you survive."

- General George S. Patton

Part 1. Preview

A. Introduction

Normally, you will spend more time moving than fighting. The fundamentals of movement discussed in this chapter provide techniques that all Soldiers must learn.

Even seasoned troops[01] should practice these techniques regularly, until they become second nature[02] .

01 seasoned troops : 정예 부대
02 second nature : 제2의 천성, 몸에 배다

B. Key Doctrinal Terms & Definition

low crawl 낮은 포복	A tactical movement technique used in military or tactical operations to move slowly and deliberately while maintaining the lowest possible profile to avoid detection or enemy fire.
high crawl 높은 포복	A tactical movement technique used in military or tactical operations to move quickly while maintaining a low profile and some cover. The high crawl lets the Soldier move faster than the low crawl while still providing the Soldier with a low silhouette.
rush 약진	A tactical movement technique used to cover open ground quickly while minimizing exposure to enemy fire. The rush is the fastest way to move from one covered position to another.
traveling 연속 전진	A movement technique used when speed is necessary and contact with enemy forces is not likely.
traveling overwatch 선두 감시하 전진	A movement technique used when contact with enemy forces is possible.
bounding overwatch 교대 전진	A movement technique used when contact with enemy forces is expected.
arm-and-hand signals 완수신호	A form of nonverbal communication that use gestures to convey a message

movement formations 전투 대형	An ordered arrangement of forces for a specific purpose and describes the general configuration of a unit on the ground.
wedge formation 삼각대(설대) 대형	A movement formation with one lead element and the trail elements are paired off abreast of each other on the flanks.

Part 2. Reading

A. Individual Movement Techniques

Your leaders base their selection of a particular movement technique[03] by traveling[04], traveling overwatch[05], or bounding overwatch[06] on the likelihood of enemy contact and the requirement for speed. However, your unit is ability to move depends on your movement skills and those of your fellow Soldiers. Use the following techniques to avoid being seen or heard:

(1) Stop, look, listen, and smell (SLLS) before moving. Look for your next position before leaving a position.

(2) Look for covered and concealed routes[07] on which to move.

(3) Change direction slightly from time-to-time[08] when moving through tall grass.

(4) Stop, look, and listen when birds or animals are alarmed (the enemy may be nearby).

(5) Avoid steep slopes[09] and places with loose dirt[10] or stones.

In addition to walking, you may move in one of three other methods known as individual movement techniques (IMT) — low crawl[11], high crawl[12], or rush[13].

03 movement technique : 이동기술
04 traveling : 연속 전진
05 traveling overwatch : 선두 감시하 전진
06 bounding overwatch : 교대 전진
07 covered and concealed routes : 엄폐 및 은폐된 통로
08 from time-to-time : 종종, 때때로
09 steep slopes : 가파른 언덕
10 loose dirt : 푸석푸석한 흙
11 low crawl : 낮은 포복
12 high crawl : 높은 포복
13 rush : 약진
14 silhouette : 윤곽
15 cover : 엄폐
16 concealment : 은폐

B. Low Crawl

The low crawl gives you the lowest silhouette[14]. Use it to cross places where the cover[15] and/or concealment[16] are very low and enemy fire or observation prevents you from getting up. Keep your body flat against the ground. With your firing hand[17], grasp[18] your weapon sling[19] at the upper sling swivel[20]. Let the front hand guard rest on your forearm (keeping the muzzle[21] off the ground), and let the weapon butt[22] drag on the ground. To move, push your arms forward and pull your firing side leg forward. Then pull with your arms and push with your leg. Continue this throughout the move (Figure 8-1).

17 firing hand : 사격할 때 사용하는 손 (ex, 오른손잡이는 오른손)
18 grasp : 붙잡다, 움켜쥐다
19 sling : 멜빵
20 sling swivel : 멜빵 고리
21 muzzle : 총구
22 weapon butt : 개머리판
23 cradle : 안듯이 잡다
24 buttocks : 엉덩이

C. High Crawl

The high crawl lets you move faster than the low crawl and still gives you a low silhouette. Use this crawl when there is good cover and concealment but enemy fire prevents you from getting up. Keep your body off the ground and resting on your forearms and lower legs. Cradle[23] your weapon in your arms and keep its muzzle off the ground. Keep your knees well behind your buttocks[24] so your body will stay low.

To move, alternately advance your right elbow and left knee, then your left elbow and right knee (Figure 8-1).

<Figure 8-1. Low and high crawl>

D. Rush

The rush is the fastest way to move from one position to another (Figure 8-2). Each rush should last from 3 to 5 seconds. Rushes are kept short to prevent enemy machine gunners[25] or riflemen[26] from tracking you. However, do not stop and hit the ground in the open just, because 5 seconds have passed. Always try to hit the ground behind some cover. Before moving, pick out your next covered and concealed position and the best route to it. Make your move from the prone position[27] as follows:

25 machine gunners : 기관총 사수

26 riflemen : 소총수

27 prone position : 엎드린 자세

(1) Slowly raise your head and pick your next position and the route to it.

(2) Slowly lower your head.

(3) Draw your arms into your body (keeping your elbows in).

(4) Pull your right leg forward.

(5) Raise your body by straightening your arms.

(6) Get up quickly.

(7) Rush to the next position.

<Figure 8-2. Rush.>

E. Movement Within A Team

Movement formations[28] are used for control, security, and flexibility. These formations are the actual arrangements for you and your fellow Soldiers in relation to each other.

Every squad and Soldier has a standard position. You must be able to see your fire team leader. Fire team leaders must be able to see their squad leaders. Leaders control their units using arm-and-hand signals[29].

Formations also provide 360-degree security and allow the weight of their firepower to the flanks or front in anticipation of enemy contact.

Formations do not demand parade ground[30] precision. Your leaders must retain[31] the flexibility needed to vary their formations to the situation. The use of formations allows you to execute battle drills[32] more quickly and gives the assurance that your leaders and buddy team members are in their expected positions and performing the right tasks. You will usually move as a member of a squad/team.

28 movement formations : 전투대형(= combat formation)

29 arm-and-hand signals : 완수신호

30 parade ground : 행진

31 retain : 유지하다

32 battle drills : 전투 훈련

<Figure 8-3. Fire team wedge formation>[33]

F. Movement in Urban Areas

Movement in urban areas is the first skill you must master. Movement techniques must be practiced until they become second nature. To reduce exposure to enemy fire, you should avoid open areas, avoid silhouetting yourself, and select your next covered position before movement. The following paragraphs discuss how to move in urban areas:

(1) AVOIDING OPEN AREAS : Open areas, such as streets, alleys, and parks, should be avoided. They are natural kill zones for enemy, crew-served weapons[34], or snipers. They can be crossed safely if the individual applies certain fundamentals, including using smoke[35] from hand grenades or smoke pots[36] to conceal movement. When employing smoke as an obscurant[37], keep in mind that thermal sighting systems can see through smoke. Also, when smoke has been thrown in an open area, the enemy may choose to engage with suppressive fires[38] into the smoke cloud.

(2) MOVING PARALLEL TO BUILDINGS : You may not always be able to use the inside of buildings as routes of advance and must move on the outside of the buildings. Smoke, suppressive fires, and cover and concealment should be used as much as possible to hide movement. You should move parallel to the side of the building, maintaining at least 12 inches of separation between yourself and the wall to avoid rabbit rounds (ricochets[39] and rubbing or bumping the wall). Stay in the shadows, present a low silhouette, and move rapidly to your next position. If an enemy gunner inside the building fires, he exposes himself to fire from other squad members providing overwatch.

33 wedge formation : 삼각대(설대) 대형

34 crew-served weapon : 공용 화기

35 smoke : 연막

36 smoke pots : 발연통

37 obscurant : 차폐제

38 suppressive fires : 제압사격

39 ricochets : 도비탄, 튕겨 나온 총알

(3) MOVING PAST WINDOWS : Windows present another hazard to the Soldier. The most common mistakes are exposing the head in a first-floor window and not being aware of basement windows. When using the correct technique for passing a first-floor window, you must stay below the window level and near the side of the building (Figure 8-4). Ensure you do not silhouette yourself in the window. An enemy gunner inside the building would have to expose himself to covering fires if he tries to engage you.

<Figure 8-4. Soldier moving past windows>

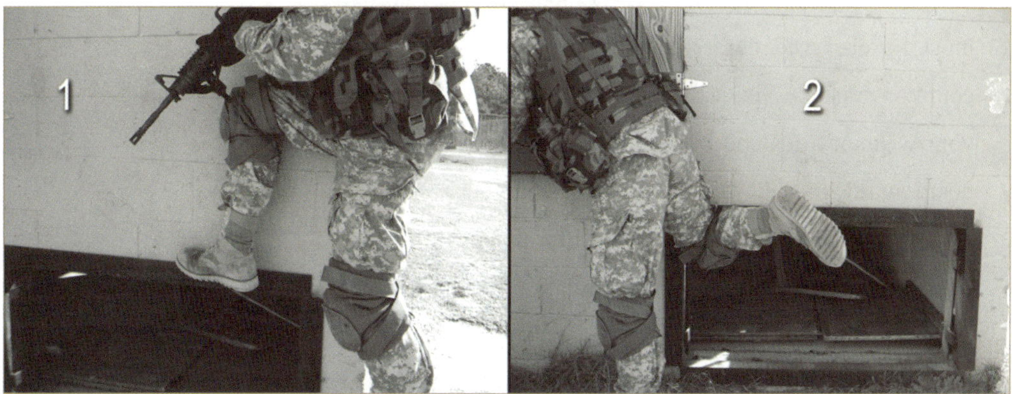

<Figure 8-5. Soldier passing basement windows>

The same techniques used in passing first-floor windows are used when passing basement windows. You should not walk or run pass a basement window, as this will present a good target for an enemy gunner inside the building. Ensure you stay close to the wall of the building and step or jump pass the window without exposing your legs (Figure 8-5).

(4) CROSSING A WALL : You must learn the correct method of crossing a wall (Figure 8-6). After you have reconnoitered[40] the other side, quickly roll over the wall and keep a low silhouette. Your speed of movement and low silhouette denies the enemy a good target.

<Figure 8-6. Soldier crossing a wall>

(5) MOVING AROUND CORNERS : The area around a corner must be observed before the Soldier moves. The most common mistake you can make at a corner is allowing your weapon to extend beyond the corner, exposing your position; this mistake is known as flagging your weapon[41]. You should show your head below the height an enemy would expect to see it. You must lie flat on the ground and not extend your weapon beyond the corner of the building. Only expose your head (at ground level) enough to permit observation (Figure 8-7). You can also use a mirror, if available, to look around the corner. Another corner-clearing technique that is used when speed is required

40 reconnoiter : 정찰하다

41 flagging your weapon : 무기를 부주의하게 겨누는 것

is the pie-ing method. This procedure is done by aiming the weapon beyond the corner into the direction of travel (without flagging) and side-stepping around the corner in a circular fashion with the muzzle as the pivot point (Figure 8-8).

<Figure 8-7. Correct technique for looking around a corner>

<Figure 8-8. Pie-ing a corner>

(6) MOVING WITHIN A BUILDING : Once you have entered a building follow these procedures to move around in it:

- Doors and Windows : Avoid silhouetting yourself in doors and windows (Figure 8-9).
- Hallways : When moving in hallways, never move alone- always move with at least one other Soldier for security.
- Walls : You should try to stay 12 to 18 inches away from walls when moving; rubbing against walls may alert an

enemy on the other side, or, if engaged by an enemy, ricochet rounds tend to travel parallel to a wall.

<Figure 8-9. Movement within a building>

Part 3. Dialogue

[Dialogue] The instructor and cadet are having a conversation during training.

INSTRUCTOR: I told you last time, in addition to walking there are three methods known as individual movement techniques. Do you remember what they are?

CADET: Yes, I do! The three main individual movement techniques are low crawl, high crawl, and rush.

INSTRUCTOR: Very nice! So, Can you explain how to perform a low crawl?

CADET: First of all, lie my body flat against the ground. And then, Keep my head low, facing forward, and rely on peripheral vision to stay aware of surroundings. Extend one arm forward, bent slightly at the elbow, to pull myself forward. And the opposite arm remains by my side. At the same time, keep my legs spread slightly apart, with one leg bent at the knee and the other extended for balance.
Use the toes for propulsion, pushing myself forward while keeping body flat. Pull with forward arm while simultaneously pushing with the opposite leg. It'll be easier to explain by showing it to you.

(The cadet lay on the ground and performed a low crawl)

INSTRUCTOR: You know exactly how to do it.

CADET: However, Moving in a low crawl is too slow and difficult. When I use this method?

INSTRUCTOR: You use a low crawl in situations where maximum concealment is necessary, and the terrain or environment offers little to no cover. The low crawl minimizes your exposure by keeping your body flat against the ground.

CADET: You mean while the low crawl is slower than other techniques, its stealth benefits make it the best choice when staying hidden is more important than speed, right?

INSTRUCTOR: You are correct. maximizes concealment, increases safety, and minimizes noise are the main reasons for a low crawl!

Part 4. Quiz

A. Multiple-Choice Questions

01. Which of the following is not an action of a low crawl?

a. Crawl using forearms to pull the body forward.

b. Use toes for propulsion while keeping legs close to the ground.

c. Crawl using both forearms and knees for propulsion.

d. Slow and deliberate to reduce noise and visibility.

02. If you want to cross open area quickly, which one is a best method for moving?

a. Walk
b. Low crawl
c. High crawl
d. Rush

03. When under observation or in exposed terrain, which one is a best method for moving?

a. Walk
b. Low crawl
c. High crawl
d. Rush

04. You need both speed and concealment and some cover is available, which one is a best method for moving?

a. Walk
b. Low crawl
c. High crawl
d. Rush

B. True/False Questions

01. The low crawl gives you the highest silhouette. (T / F)

02. The high crawl lets you move faster than the low crawl and still gives you a low silhouette. (T / F)

03. Each rush should last from 3 to 5 seconds. (T / F)

04. Formations also provide 180-degree security and allow the weight of their firepower to the flanks or front in anticipation of enemy contact. (T / F)

05. The low crawl is for maximum concealment. Used when avoiding detection is critical, and the terrain provides little or no cover. (T / F)

06. The high crawl is the fastest of the three techniques. (T / F)

07. The advantages of a rush are covering ground quickly and reducing exposure time. (T / F)

08. Use the low crawl when staying hidden is the priority, especially under enemy observation. (T / F)

09. Use the high crawl when you need to move faster and can afford to sacrifice some concealment due to available cover or reduced threat. (T / F)

10. Formations demand parade ground precision. (T / F)

C. Fill-in-the-Blan k Questions

01. In addition to walking, you may move in one of three other methods known as individual movement techniques (IMT) — _________, _________, or _________.

02. The __________ gives you the lowest silhouette.

03. The __________ lets you move faster than the low crawl and still gives you a low silhouette.

04. The __________ is the fastest way to move from one position to another

05. Movement formations are used for control, security, and __________.

06. Leaders control their units using __________ signals.

07. To do a rush, make your move from the __________ as follows:

- Slowly raise your head and pick your __________ and the route to it.
- Slowly lower your head.
- Draw your arms into your body (keeping your elbows in).
- Pull your __________ forward.
- Raise your body by straightening your arms.
- Get up quickly.
- __________ to the next position.

Army stories

워리어 플랫폼 (Warrior Platform)

워리어 플랫폼이란 육군의 기본 전투 요소인 각개 전투원(Warrior)이 전투력 발휘를 위해 착용하는 피복 · 장구 · 장비로 구성된 기반체계 (Platform)을 말한다. 워리어 플랫폼 도입에 따라 우리 군에서 기존에 적용하던 약진, 포복 등의 이동 기술이 일부 개선되고 있다. 이에 우선 워리어 플랫폼에는 어떤 것들이 있고, 또 이를 착용한 각개 전투원들의 형태가 어떻게 바뀌게 되는지 알아보자.

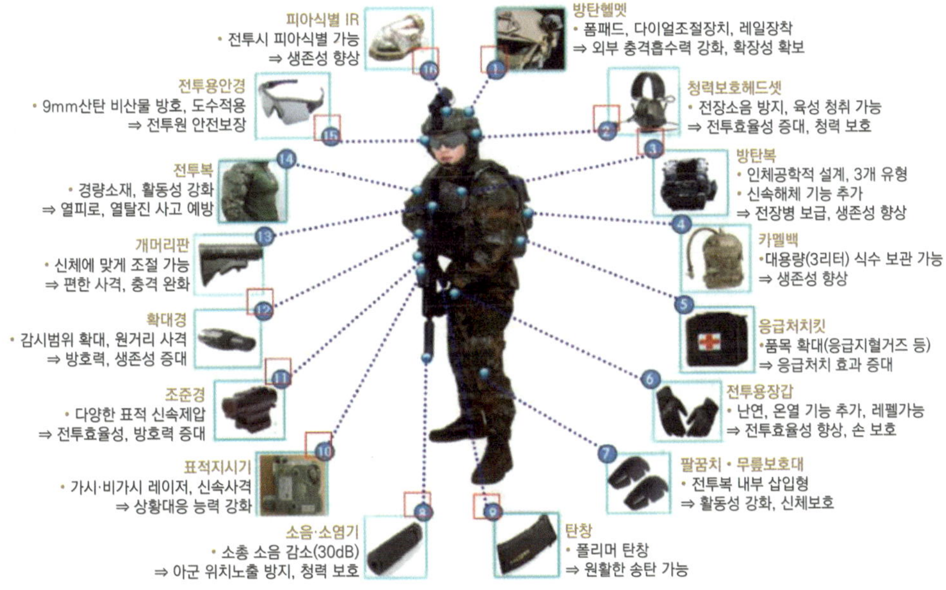

피복류에서부터 장구류까지 총 35종의 품목들이 개선 또는 신규 전력화하게 되며, 26년까지 전력화 시키는 것을 목표로 단계별로 추진중에 있다.

<table>
<tr><th colspan="3">구분</th><th>1단계['23]</th><th>2단계('25년]</th><th>3단계('26년 이후)</th></tr>
<tr><td colspan="6">「워리어플랫폼」대상품목</td></tr>
<tr><td rowspan="3">전력지원체계</td><td>피복</td><td>①전투복 ② 전투화
③방상내, 외피 ④전투우의
⑤전투용장갑 ⑥ 내의류
⑦ 방한화 ⑧기능성방한복
⑨ 모자류 ⑩방한소품</td><td>계속사업
(중기반영)
- 품질개선, 성능개량
-대상 : 36.5만명</td><td>계속사업
-품질개선, 성능개량
-대상 : 36.5만명</td><td>계속사업
-품질개선, 성능개량
-대상 : 36.5만명</td></tr>
<tr><td>장구</td><td>① 방탄복 (I,II,III)
② 방탄헬멧
③ 전투조끼 ④ 응급처치키트
⑤ 전투용배낭
⑥ 전투용안경 구
⑦ 수통(카멜백) ⑧ 개인천막
⑨ 침낭 ⑩ 야전삽</td><td>계속사업
(중기반영)
-품질개선, 성능개량
-대상 : 36.5만명</td><td>계속사업
-품질개선, 성능개량
-대상 : 36.5만명</td><td>계속사업
-품질개선, 성능개량
-대상 : 36.5만명</td></tr>
<tr><td>장비</td><td>① 청력보호헤드셋
② 피아식별 IR
③ 대검 ④ 탄창
⑤ 표적지시기 ⑥ 조준경
⑦ 확대경 ⑧ 소음. 소염기
⑨ 원거리조준경</td><td>신규 반영
(추가 예산)
-대상 : 4.4만명
-예산: 000억원</td><td colspan="2">기술 수준을 고려
개인전투체계 Block-1과
일부품목 통합 검토</td></tr>
<tr><td rowspan="2">무기체계</td><td rowspan="2">장비</td><td>① 개인화기 ② 야간투시경</td><td colspan="3">성능개량 등 소요발생시 방위력 개선사업 추진</td></tr>
<tr><td>① 정보입력기 ② 정보처리기
③ 영상전시기 ④ 개인무전기
⑤ 통합전원
⑥ 생체리듬확인체계</td><td>*개인전투체계 Block-I 연구개발
*'19년 연구개발비 40억원 기재부 요구 (방시청)</td><td colspan="2">*개인전투체계 Block-I 연구개발</td></tr>
<tr><td colspan="6">「개인전투체계 사업」대상품목</td></tr>
<tr><td colspan="6">※네트워크 환경(NCOE)하 개인의 감시 / 타격능력, 생존성을 향상시키기 위해 6개품목 000세트를 확보하는 사업</td></tr>
</table>

〈워리어 플랫폼 대상 품목 및 전력화 계획〉

이를 하나씩 살펴보면, 우선 방탄헬멧과 방탄복의 개선이다. 방탄헬멧은 방호능력은 강화하고 인체공학적 디자인을 도입하여 전투하중 및 전투피로도를 감소시켰다. 방탄복은 적 위협강도에 따라 모듈화하여 착용할 수 있도록 개선된다.

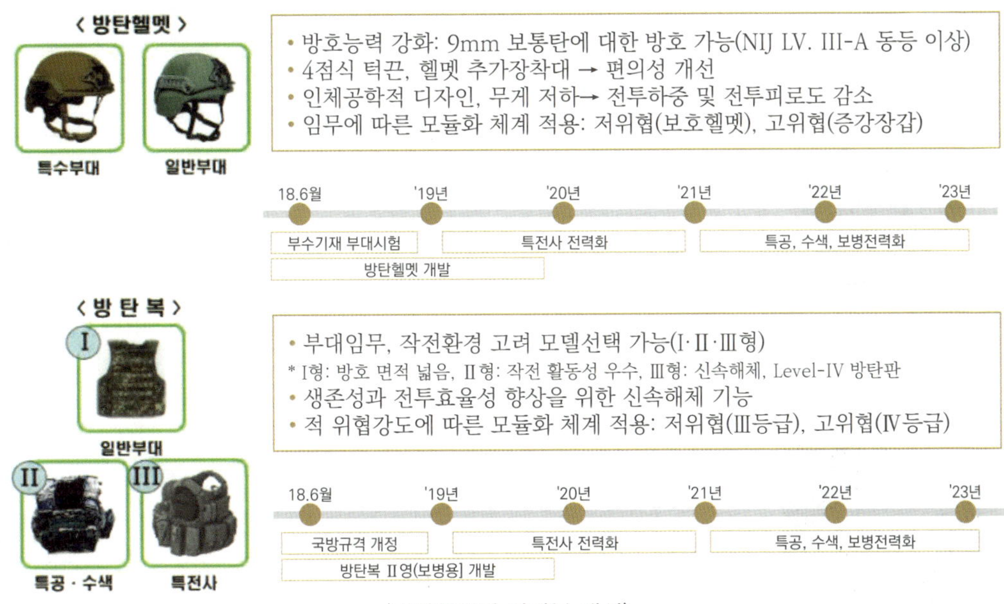

〈방탄헬멧과 방탄복 개선〉

그리고 청력보호 및 통신을 용이하게 하기 위해 청력보호 헤드셋과 시력 보호 및 효율적 시각화를 위한 전투용 안경이 도입된다.

< 청력보호해드셋 >

• 총성, 폭음 등 최대 30db로 감소(총성→ 일반적 대화 시 목소리)
• 작은 대화소리는 증폭시키며, 스테레오 방식으로 발성위치 판단 가능
• 개인 무전기와 연동 가능한 송수신장비(PPT)를 활용한 다채널 통신
• 악천후 시 운용이 가능한 완전방수(IP68) 등급의 내구성 확보

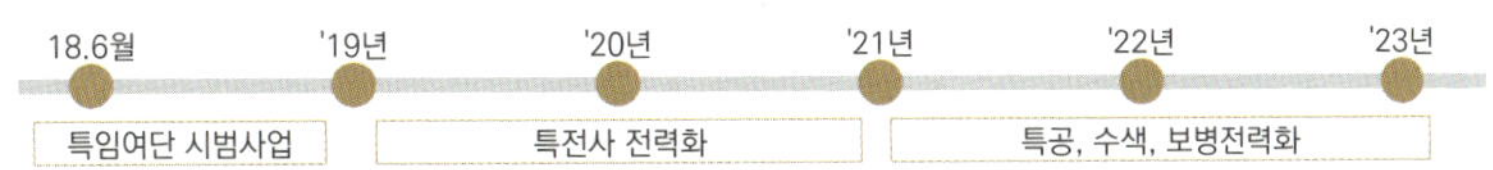

< 전투용안경 >

• 美 군사규격(MILPRF 32432)을 충족하는 폴리카보네이트 사용(파편 방호)
• 작전 및 임무특성을 고려한 신속한 렌즈교체(투명·검정 등)
• 자외선 및 레이져에 대한 안구 보호기능 강화
• 동양인 두상 및 얼굴 형상을 고려한 인체공학적 설계

〈청력보호헤드셋과 전투용 안경 도입〉

또한 효율적 전투를 위한 조준경, 레이져표적지시기, 고성능 확대경, 소음/소염기, 피아식별 IR 같은 장비도 대거 개선 및 도입된다.

〈조 준 경〉

• 조준점(Red dot) 및 Parallax-free(親差 의 "0화")로 정확한 조준사격 가능
• 기존 PVS-11K (주야간조준경) 대비 운용시간 증가
• QD(Quick Detach) 레버 활용 탈착 용이, 일정압력으로 영점유지 가능
• 임무 및 작전특성에 따른 조준점 선택: 조준점 형태(•, ⊙, ⊙ 등)

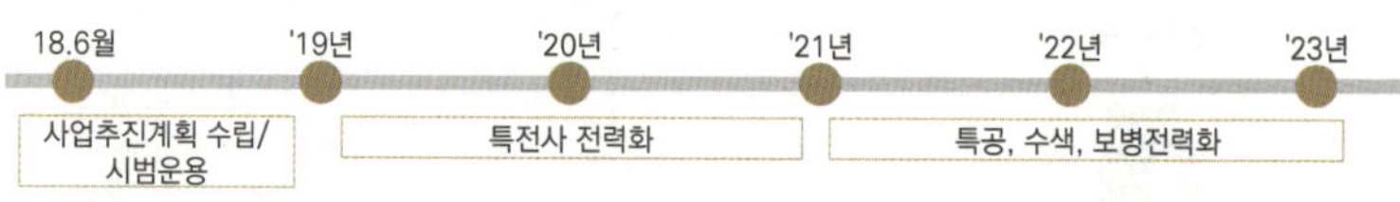

〈 레이져표적지시기 〉

• 주·야간 모두 활용 가능(가시. 비가시 레이져로 구성)
• 향후 실전 훈련 모드 적용(실전 : 고출력, 훈련: 저출력, 안구 망막 보호
• 야투경과 통합운용 시 원활한 야간작전 수행 가능(적외선 탐조등 기능)
• 포병, 박격포, 공중화력에 대한 표적지시 가능(Beam 조사거리 : 2~3km)

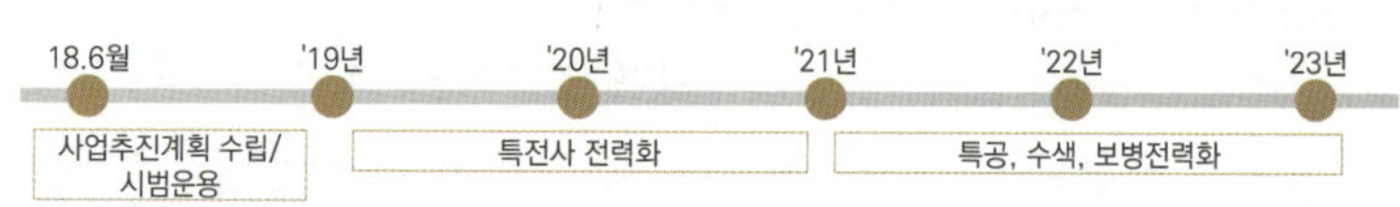

〈조준경과 레이져표적지시기 〉

이러한 장비들의 도입으로 전투효율성이 극대화되고, 개인 전투원들의 생존성은 향상될 것으로 기대되고 있다. 첨단과학기술이 군에 도입됨으로 인해 군의 전투력 강화는 물론 국내 방산기술 역량도 높아질 것이다. 더불어, 최첨단 장비 및 물자를 보급받은 장병들의 사기와 자긍심도 한층 더 배양될 것이라 예상된다.

09

TACTICS

Fire team, Squad, & Platoon

By the end of this chapter, you will be able to

- name the duty positions that make up a fire team & squad.
- identify the weapons that each member carries.
- understand the organization and role of a fire team & squad.

"I love the infantry because they are the underdogs. They are the mud-rain-frost-and-wind boys. They have no comforts, and they even learn to live without the necessities. And in the end they are the guys that wars can't be won without."

*- Ernie Pyle
(World War II correspondent)*

Part 1. Preview

A. Introduction

This section explains the elements, weapons, roles, and responsibilities of the fire team and rifle squad. Success in tactical[01] operations depends on Soldiers at all levels understanding their tactical mission and the steps necessary to accomplish it.

01 tactical : 전술의

B. Key Doctrinal Terms & Definition

command and control 지휘통제	The exercise of authority and direction by a properly designated commander over assigned and attached forces in the accomplishment of the mission.
maneuver 기동	Employment of forces in the operational area, through movement in combination with fires and information, to achieve a position of advantage in respect to the enemy.
fire and movement 사격과 기동	The concept of applying fires from all sources to suppress, neutralize, or destroy the enemy, and the tactical movement of combat forces in relation to the enemy
troop leading procedure 소부대지휘절차	A dynamic process used by small-unit leaders to analyze a mission, develop a plan, and prepare for an operation.
commander's Intent 지휘관 의도	A clear and concise expression of the purpose of the operation and the desired military end state that supports mission command, provides focus to the staff, and helps subordinate and supporting commanders act to achieve the commander's desired results without further orders, even when the operation does not unfold as planned.
movement formation 전투대형	An ordered arrangement of forces for a specific purpose and describes the general configuration of a unit on the ground.

METT-TC(I) 임무변수	The mission variables leaders use to analyze and understand a situation in relationship to the unit's mission. *mission, enemy, terrain and weather, troops and support available, time available, civil considerations, and informational considerations
direct fires 직접화력	Fire delivered on a target using the target itself as a point of aim for either the weapon or the director.
suppressive fire 제압사격	Fires on or about a weapons system to degrade its performance below the level needed to fulfill its mission objectives during the conduct of the fires.
urban operations 도시지역작전	Operations across the range of military operations planned and conducted on, or against objectives on a topographical complex and its adjacent natural terrain, where man-made construction or the density of population are the dominant features.

Part 2. Reading

A. Maneuver element of infantry small unit

The Army's smallest maneuver element controlled by a leader is the fire team. The fire team is the building block for all Army tactical operations. Fire teams make up squads; squads make up platoons. Army lieutenants lead platoons as part of an infantry rifle company.

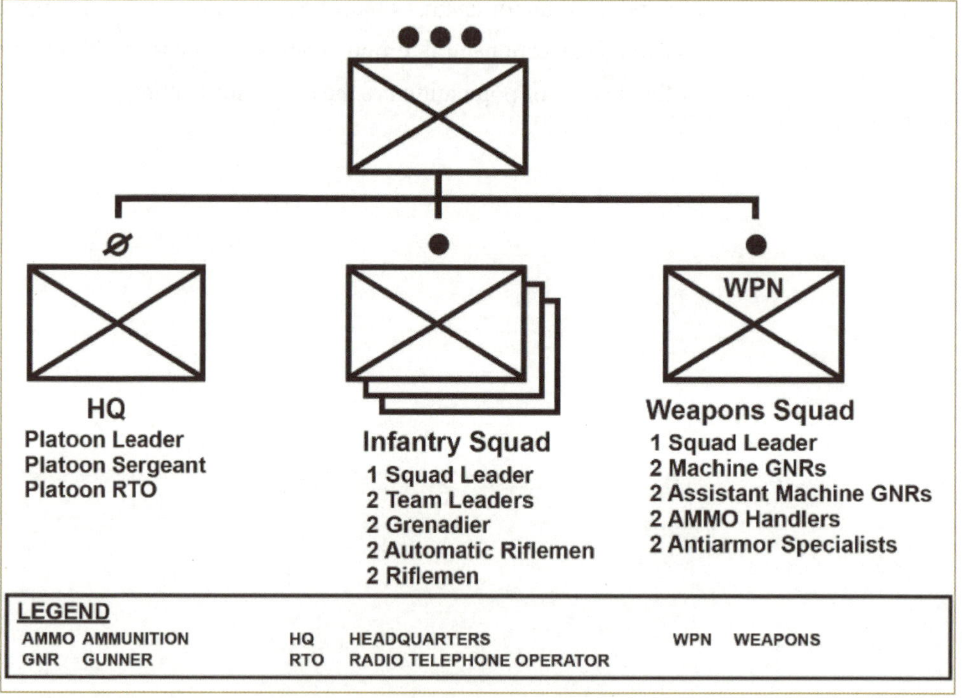

<Organization of Infantry Rifle platoon>

The infantry is often described as "the tip of the spear." The four Soldiers in a fire team and the nine Soldiers in a rifle squad represent the sharpened tip of that spear. Success in a mission depends on all Soldiers in the fire team and squad understanding the unit's mission—its task and purpose—

and how each Soldier's assigned weapon and each Soldier's role and responsibility relate to mission success. Fire teams and squads led by the Army lieutenant form the Army's basic maneuver[02] element—the infantry rifle platoon—the foundation on which all tactical operations rest.

B. The Elements of a Fire Team

A squad's fire teams are referred to as the Alpha Team(A team) and Bravo Team(B team). Each fire team has four Soldiers—a fire team leader, a rifleman, an automatic rifleman, and a grenadier. The team members' rank and experience will range from a private (E-1) straight out of Initial Entry Training(IET) [03] and Advanced Individual Training(AIT) to a specialist (E-4) who may have anywhere from one to four years of experience. The team leader is generally a sergeant(E-5) with three to five years of experience.

The team members' positions[04] (rifleman[05], automatic rifleman[06], and grenadier[07]) are based on their assigned weapons (rifle, squad automatic weapon [SAW], and M203 grenade launcher). The platoon leader and platoon sergeant consider each Soldier's experience, skill with the weapon, and longevity[08] in the unit before assigning the Soldier a position on the fire team. The rifleman, team leader, and squad leader carry rifles from the M16/M4 family of weapons, based on the table of organization and equipment(TO&E)[09] for various units.

C. The elements of Squads and Platoon

The Infantry rifle squad's primary role is as a maneuver or base-of-fire element. The squad is comprised of two fire teams

02 maneuver : 기동
03 Initial Entry Training : 신병훈련, 기초군사훈련
04 positions : 직책
05 rifleman : 소총수
06 automatic rifleman : 기관총사수
07 grenadier : 유탄수
08 longevity : 근속기간
09 table of organization and equipment : 편제표

and a squad leader. The squad leader and two team leaders provide the squad with command and control[10]. Additionally, the squad has two rifleman, two automatic rifleman, and two grenadiers. It can establish abase of fire, providing security for another element, or conducting fire and movement with one team providing a base of fire, while the other team moves to the next position of advantage or onto an objective. The squad leader has two subordinate[11] leaders to lead the two teams, freeing the squad leader to control the entire squad.

10 command and control : 지휘통제

11 subordinate : 부하, 예하의

12 drill : 훈련

13 organic : 편제의

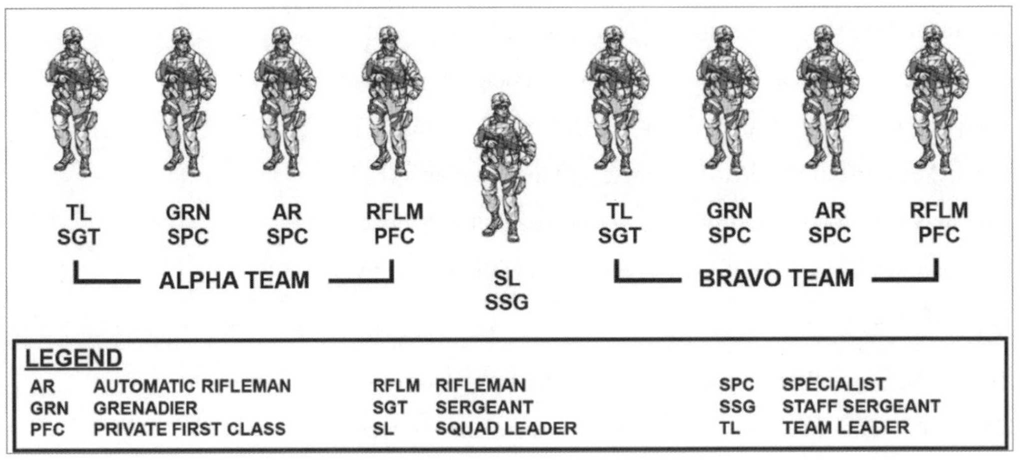

<Infantry Rifle Team and Squad>

The rifle squad leader is generally a staff sergeant(E-6) with six to eight years of experience, who came up from the ranks and started as a fire-team rifleman. The squad leader is responsible for everything the squad does or fails to do. He or she is a tactical leader and leads by example. Among other things, the squad leader:

- Acts as the subject matter expert on all squad-level battle drills[12] and individual Soldier tasks.
- Acts as the subject matter expert for the squad's organic[13] weapons employment and employment of supporting assets.

- Controls the movement of the squad and its rate and distribution of direct fires[14].
- Fights the close fight by fire and movement[15] with two fire teams and available supporting weapons.
- Communicates timely and accurate situation reports and status reports
- Ensures Soldiers maintain all equipment and account for equipment and supplies.
- Conducts Troop Leading Procedure(TLP)[16].
- Understands the mission and commander's intent[17] two levels up.

Most of the maneuver and firepower of a rifle platoon derives from the platoon's three rifle squads. Together, the rifle squads and headquarters element make up the rifle platoon an Army lieutenant leads. The platoon leader leads Soldiers by personal example and is responsible for accountability of platoon members, assigned equipment, all the platoon does or fails to do, and having complete authority over the subordinates. This centralized authority enables the platoon leader to maintain unit discipline[18], unity, and to act decisively. The platoon leader must be prepared to exercise initiative[19] within the company commander's intent and without specific guidance for every situation. The platoon leader knows the Soldiers of the platoon, how to employ the platoon, its weapons, and its systems. During operations, the platoon leader—

- Leads the platoon in supporting the higher headquarters missions.
- Conducts troop leading procedures (TLP).

14 direct fires : 직접화력
15 fire and movement : 사격과 기동
16 Troop Leading Procedure(TLP) : 소부대 지휘절차
17 commander's intent : 지휘관 의도
18 discipline : 규율, 훈련
19 initiative : 주도권

- Maneuvers squads and fighting elements and synchronizes the efforts of squads.
- Plans and controls employment of direct and indirect fires[20].
- Employs command and control systems available to the squads and platoon.
- Assigns clear tasks and purposes to the squads.
- Understands the mission and commander's intent two levels up.(CO and BN)

A platoon leader exercises command and control as an integral part of the headquarters section of the platoon. The headquarters section consists of the platoon leader, the platoon sergeant, and the platoon radiotelephone operator[21] (known as RTO), and it provides the platoon with command and control, and communications.

D. Role of the infantry rifle platoon and squads

The primary mission of the Infantry rifle platoon and squad is to close with[22] the enemy by means of fire and movement[23] to destroy, capture, or repel an assault by fire, close combat, and counterattack. At the squad level, fire and maneuver entails a team placing suppressive fire[24] on the enemy as another team moves against or around the enemy. The inherent strategic mobility of Infantry units dictates a need to be prepared for rapid deployment in response to situations in different operational environments. The Infantry rifle platoon and squad within the Infantry rifle company can deploy rapidly and it conducts operations against conventional[25] and unconventional enemy forces in all types of terrain and climate conditions.

20 indirect fire : 간접화력
21 radiotelephone operator : 통신병
22 close with : 교전하다
23 fire and movement : 사격과 기동
24 suppressive fire : 제압사격
25 conventional : 전통적인, 재래식의

The platoon's composition and training uniquely equip it to conduct its mission. In addition to its primary warfighting missions, the rifle platoon and squad may be tasked to perform other types of operations semi-independently or as an integral part of a larger force.

E. Movement formations

A movement formation[26] is an ordered arrangement of forces for a specific purpose and describes the general configuration of a unit on the ground. They are composed of two variables: lateral frontage, represented by the line formation[27]; and depth, represented by the column formation[28]. Leaders combine the elements of lateral frontage and depth to determine the best formation for their situation. In addition to the line and column/file, the other five types of formations—box, vee, wedge, diamond, and echelon—combine these elements to varying degrees.

Fire Team Movement Formations

The most common fire team formations among the movement options are the fire team wedge[29] and fire team column/file[30]. The wedge is the basic formation of the fire team. The interval[31] between Soldiers in the wedge formation is normally 10 meters. The wedge expands and contracts depending on the terrain. The normal interval is adjusted according to operational environment so all team members still can see their team leader and all team leaders still can see their squad leader. Team leaders use the column/file when employing the wedge is impractical. This formation most often is used in severely restrictive terrain[32], when inside a building, in dense vegetation, experiencing limited visibility[33], and so forth.

26 movement formation : 전투대형
27 line formation : 횡대
28 column formation : 종대
29 wedge : 쐐기
30 column/file : 열/종대
31 interval : 간격
32 terrain : 지형
33 visibility : 시계(視界)

The distance between Soldiers in the column/file changes due to constraints of the situation, particularly when in urban operations[34].

34 urban operations : 도시지역작전

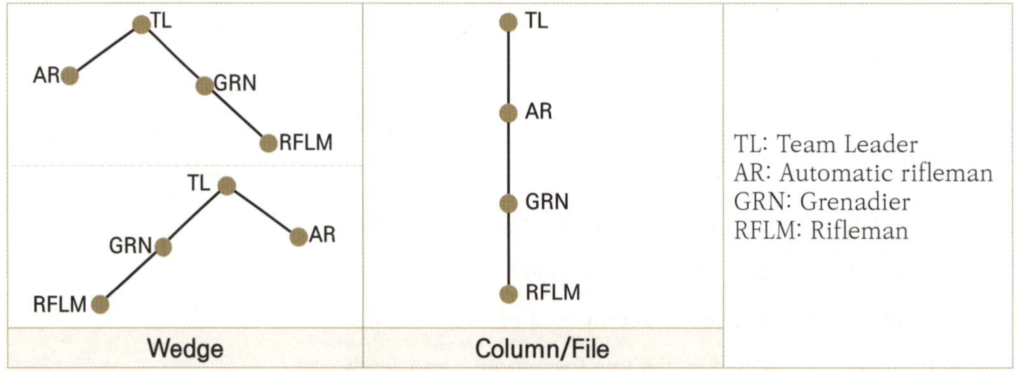

<Fire team movement formation>

Squad Movement Formations

The term squad movement formation refers to the relative locations of the fire teams. Squad formations include the squad column, the squad line, and squad file. Squad leaders adjust their squad's formation as necessary while moving, primarily through the three movement techniques, those are traveling, traveling overwatch, and bounding overwatch. Squad leaders exercise command and control primarily through the two team leaders and are responsible for 360-degree security, for ensuring the team's sectors of fire are mutually supporting, and for being able to rapidly transition the squad upon contact.

Traveling	a movement technique used when speed is necessary and contact with enemy forces is not likely.
Traveling overwatch	a movement technique used when contact with enemy forces is possible.
Bounding overwatch	a movement technique used when contact with enemy forces is expected.

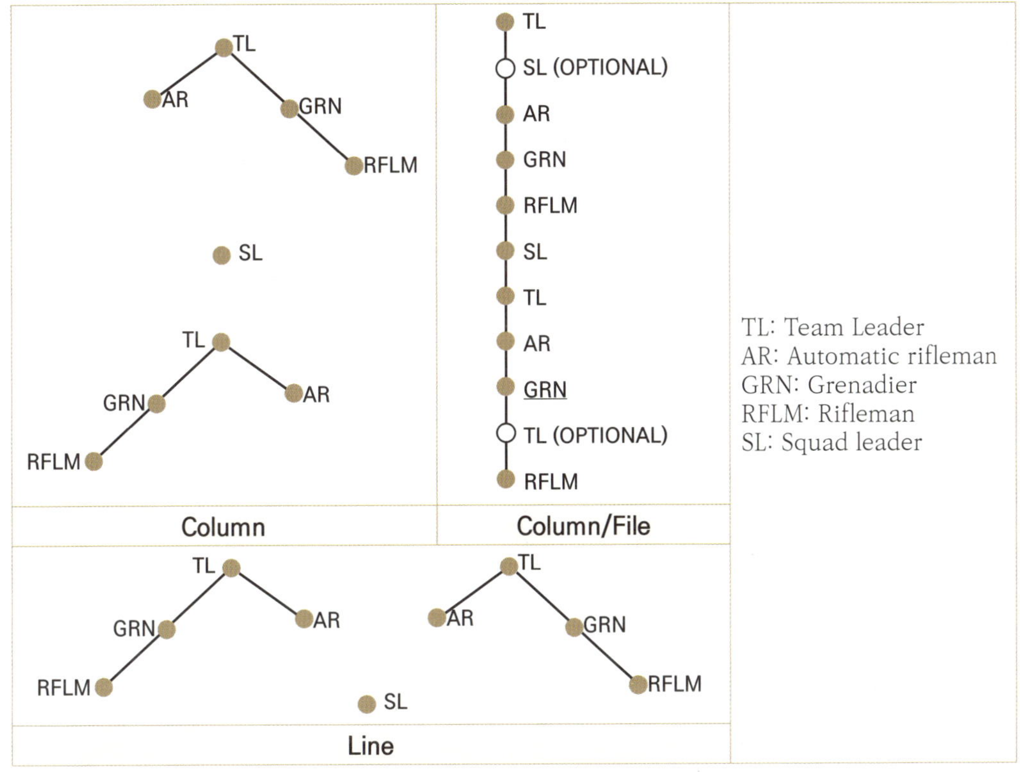

<Squad movement formation>

Platoon Movement Formations

The actual number of useful combinations of squad and fire team movement formations within the platoon movement formations is numerous, creating a significant training requirement for the unit. Infantry platoons and squads often move as part of a larger unit's movement. The next higher commander assigns the platoon a position within the formation. Platoon leaders assign their subordinates an appropriate formation based on the situation and use the appropriate movement technique. Platoon formations include the column, the line (squads online or in column), the vee, the wedge, and the file. Leaders should weigh these carefully to select the best formation based on the unit's mission and on their METT-TC(I)[35] analysis. METT-TC (I) considerations

35 METT-TC(I) : 임무변수

determine where the weapons squad or medium machine gun teams locate in the platoon formation.

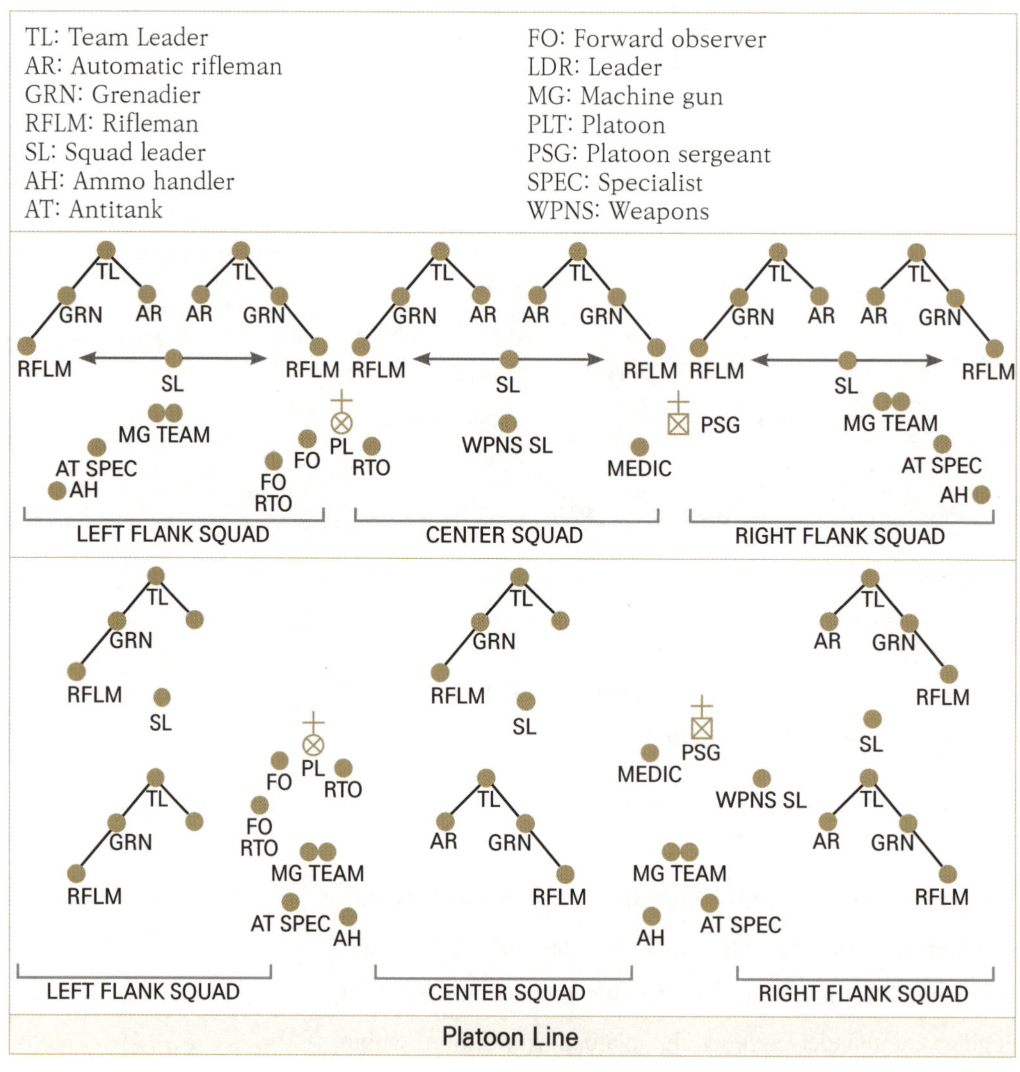

Platoon Line

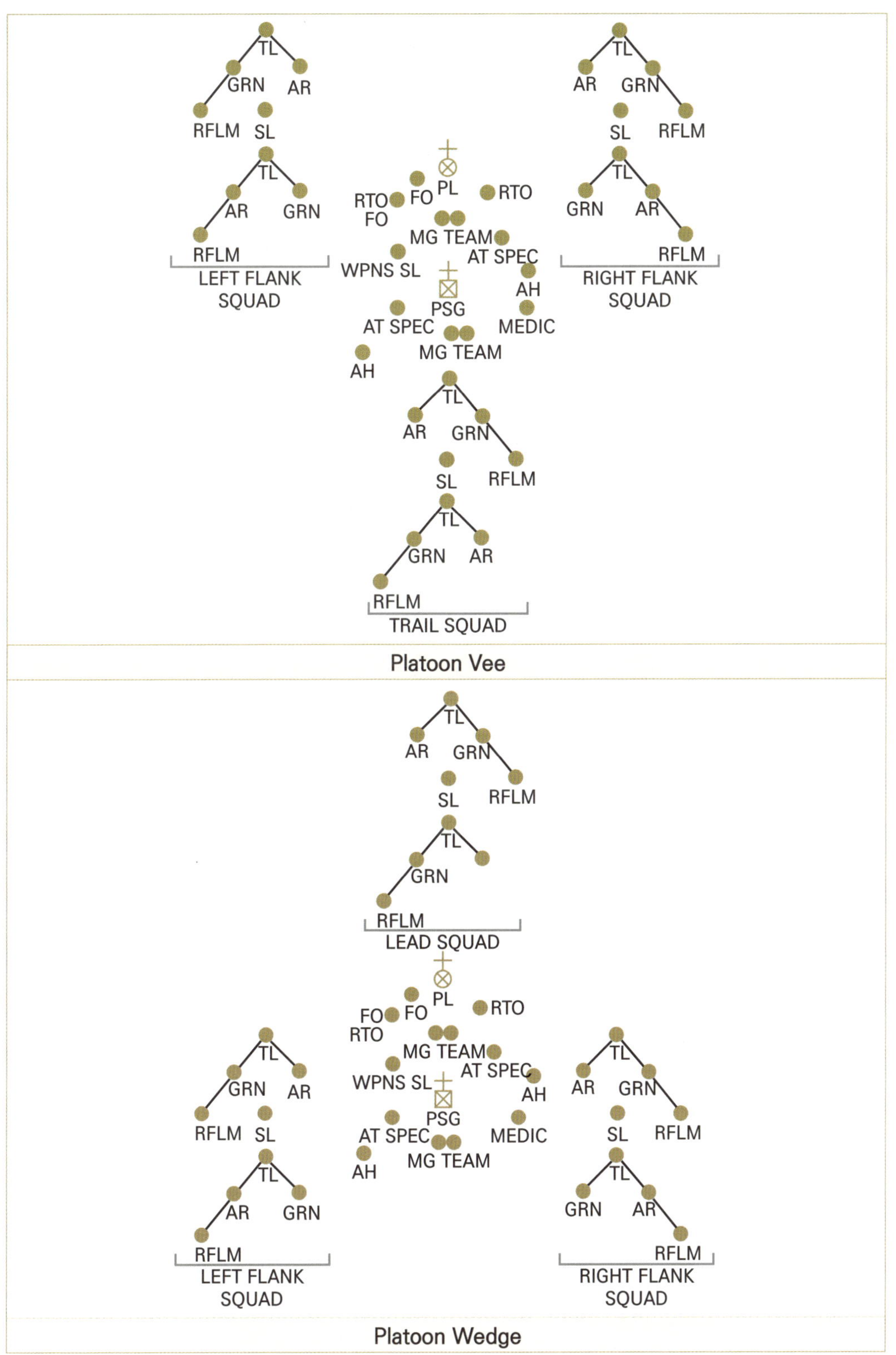

Platoon Vee

Platoon Wedge

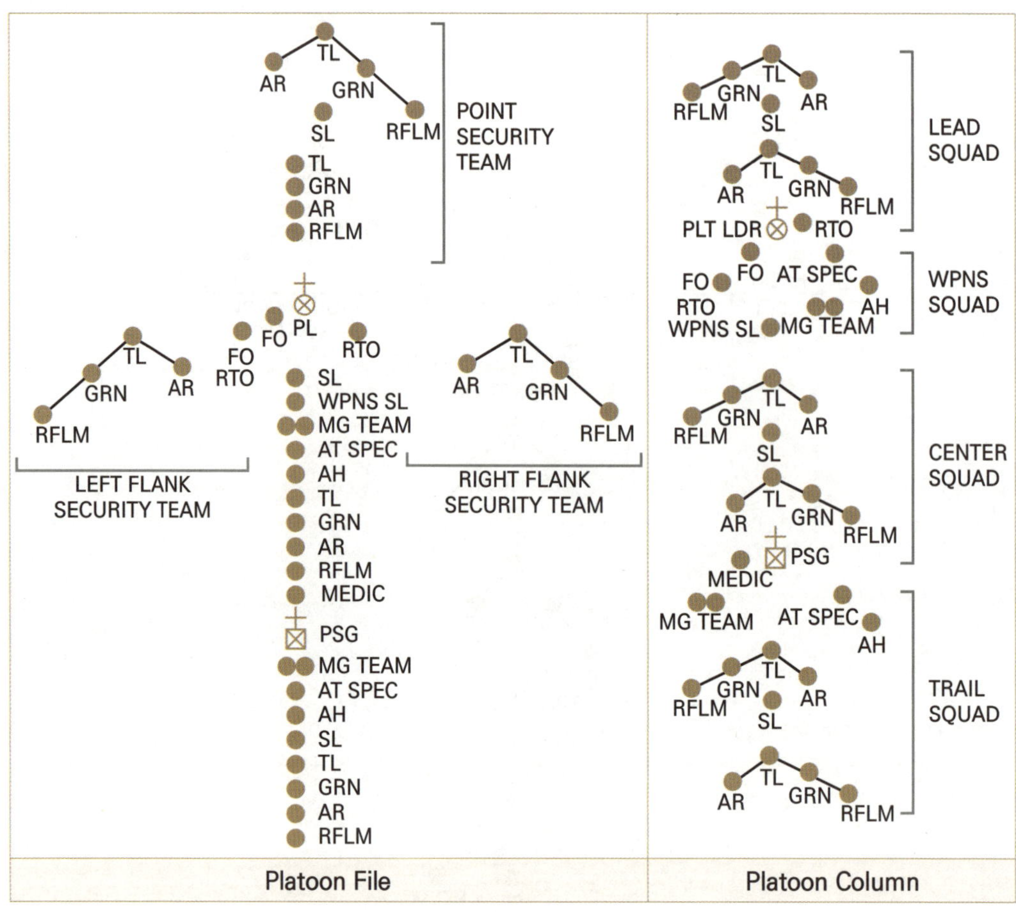

TL
AR
GRN
SL
RFLM
POINT SECURITY TEAM
TL
GRN
AR
RFLM
PL
FO
RTO
FO
RTO
TL
GRN
AR
RFLM
LEFT FLANK SECURITY TEAM
SL
WPNS SL
MG TEAM
AT SPEC
AH
TL
GRN
AR
RFLM
MEDIC
PSG
MG TEAM
AT SPEC
AH
SL
TL
GRN
AR
RFLM
TL
AR
GRN
RFLM
RIGHT FLANK SECURITY TEAM
Platoon File
TL
GRN
RFLM
SL
AR
TL
AR
GRN
RFLM
PLT LDR
RTO
LEAD SQUAD
FO
FO
AT SPEC
RTO
AH
WPNS SL
MG TEAM
WPNS SQUAD
TL
GRN
RFLM
AR
SL
TL
AR
GRN
RFLM
PSG
MEDIC
CENTER SQUAD
MG TEAM
AT SPEC
AH
TL
GRN
RFLM
AR
SL
TL
AR
GRN
RFLM
TRAIL SQUAD
Platoon Column

Part 3. Dialogue

[Dialogue 1] A platoon leader checks his squad leader on the understanding and preparation of his unit and mission.

PLATOON LEADER: Squad Leader, can you brief me on the composition of your squads?

SQUAD LEADER: Yes, Sir. Each of my squads consists of two fire teams, with four Soldiers per team. We also have a team leader for each fire team.

PLATOON LEADER: Good. Remember, the fire team is our basic maneuver element. Does every Soldier understand their role in the mission?

SQUAD LEADER: Yes, Sir. We've gone over the mission details and each Soldier knows their task and purpose. We've also reviewed how their assigned weapons and responsibilities contribute to our overall success.

PLATOON LEADER: Excellent. And what's your assessment of their readiness?

SQUAD LEADER: Sir, they're prepared. We've been focusing on communication and coordination within the fire teams to ensure they act as the "sharpened tip of the spear."

PLATOON LEADER: That's what I like to hear. Ensure the fire team leaders maintain that sharp focus. We're counting on each team to perform flawlessly when we execute.

SQUAD LEADER: Understood, Sir. I'll ensure they're ready.

[Dialogue 2] A platoon leader and his subordinates are talking about his role and mission of the platoon.

PLATOON LEADER: Private Kim, as part of your fire team, do you understand your role in today's mission?

RIFLEMAN: Yes, Sir. My task is to provide covering fire with my M4 rifle to support the team's movement during the assault phase.

PLATOON LEADER: Good. Remember, success depends on everyone knowing their role and working together. What's your fire team leader's role in this mission?

RIFLEMAN: Sir, the fire team leader coordinates our actions and ensures we stay aligned with the squad's objectives. He's also responsible for directing fire and making quick adjustments if needed.

PLATOON LEADER: That's correct. And how does your role connect to the fire team's overall mission?

RIFLEMAN: Sir, my role is to suppress the enemy and protect my teammates during movement, ensuring we can advance or hold our position effectively.

PLATOON LEADER: Excellent. Always stay alert and trust your training. The fire team is the foundation of all tactical operations, and every member plays a critical part.

RIFLEMAN: Understood, Sir. I'll do my best.

Part 4. Quiz

A. Multiple-Choice Questions

01. What is the smallest maneuver element in the Army led by a leader?

a. Fire team
b. Squad
c. Platoon
d. Infantry company

02. How many Soldiers make up a rifle squad?

a. Four
b. Nine
c. Eleven
d. Twelve

03. Who typically leads a platoon?

a. Squad leader
b. Fire team leader
c. Army lieutenant
d. Infantry company commander

04. Which of the following is often described as "the tip of the spear"?

a. Infantry rifle company
b. Fire team and squad
c. Platoon headquarters
d. Weapons team

05. What is the primary mission of the Infantry rifle platoon and squad?

a. Conducting reconnaissance missions
b. Managing logistical support
c. Closing with the enemy by fire and movement
d. Providing humanitarian assistance

06. The headquarters section of a rifle platoon typically includes which of the following members?

a. Platoon leader, squad leader, and rifleman

b. Platoon sergeant, squad leader, and radiotelephone operator

c. Platoon leader, platoon sergeant, and radiotelephone operator

d. Company commander, platoon leader, and platoon sergeant

B. True/False Questions

01. A platoon is composed of fire teams only. (T / F)

02. Each Soldier's weapon and role are crucial to the success of the mission. (T / F)

03. The fire team is the foundation of all tactical operations in the Army. (T / F)

04. The platoon leader is solely responsible for leading the rifle squads, while the platoon sergeant handles all accountability for members and equipment. (T / F)

C. Fill-in-the-Blank Questions

01. The Army lieutenant leads the infantry rifle platoon, which is the basic __________ element of the Army.

02. The fire team is composed of ______ Soldiers, while a rifle squad consists of ______ Soldiers.

03. The success of a mission depends on every Soldier in the fire team and squad understanding the unit's _______ and _______.

04. During operations, the platoon leader must understand the mission and the commander's intent ______ levels up.

E. Scenario-based Question

01. You are leading a rifle platoon and encounter an enemy force in rugged terrain. Your mission requires you to destroy the enemy position. Which of the following actions would align with the Infantry rifle platoon's role?

a. Coordinate with higher headquarters to delay the enemy until reinforcements arrive.

b. Use fire and maneuver tactics to suppress the enemy while closing with their position.

c. Conduct a retreat to avoid unnecessary casualties.

d. Set up a defensive perimeter and await further instructions.

Army stories

Origin of "Squad" 분대의 기원

"Squad"라는 용어는 프랑스어 "escouade" 또는 이탈리아어 "squadra"에서 유래했으며, 이는 소규모 방진(정사각형 또는 정렬된 집단)이나 정사각형을 의미한다. 이 용어는 르네상스 시대(15~16세기) 이탈리아 군대에서 처음 군사적 맥락으로 사용되었으며, 당시 전투에서 보다 체계적인 전술을 위해 작은 보병 부대들이 조직되었던 것이 기원이다. 이후 전쟁이 점점 더 구조화되면서 "squad"는 대형 부대 내에서 작은 전술 단위를 의미하게 되었다.

17세기에는 이탈리아어와 프랑스어의 영향을 통해 영어에 도입되었으며, 오늘날 대부분의 현대 군대에서 사용되는 표준 용어가 되었다. 현대 군대에서 분대(squad)는 8~12명의 병사로 구성된 가장 작은 전술 단위이며, 일반적으로 하사가 지휘한다.

Origin of "platoon" 소대의 기원

"Platoon"이라는 용어는 프랑스어 "peloton(pelote)"에서 유래했으며, 이는 작은 공, 덩어리 또는 소규모 집단을 의미한다. "peloton"은 공(ball)이라는 뜻의 라틴어 "pila"가 프랑스어로 전해진 것이다. 군사적 맥락에서 "peloton"은 17세기 유럽 군대에서 사용되었으며, 이 용어는 소수의 병사들이 일제 사격을 수행하는 작은 그룹을 지칭했다. 당시의 선형 전술에서는 집단적으로 사격을 조율하는 일제사격이 매우 중요했으며, 이 과정에서 소규모 부대인 "peloton"이 중요한 역할을 했다. 시간이 흐르면서 영미권으로 전해지면서 "platoon"이라는 용어로 발전하였고, 분대보다 크고 중대보다 작은 군대 조직을 의미하게 되었다. 18~19세기에 소대(platoon)는 보병 중대의 표준 하위 단위로 자리 잡았으며, 소위(Second lieutenant)와 같은 초급 장교가 지휘하게 되었다.

현대 군대에서 소대(platoon)는 일반적으로 20~50명의 병사로 구성되며, 장교(소위)가 소대장으로서 지휘를 한다.

Origin of "Lieutenant" 중·소위의 기원

임관 후 소대급 부대를 지휘하는 장교를 "lieutenant"라 하며, 프랑스어의 "lieu"와 "tenant"의 합성어에서 유래한 듯 하다. "lieu"는 장소, 현장(place)이라는 뜻이며, "tenant"는 잡다, 유지(보존)하다라는 뜻을 가진 "tenir"의 현재분사형으로 잡고있는, 유지하고있는 이라는 뜻이다. 즉, 전투현장을 직접 지휘하는 계급이라는 뜻이다. 초기에는 "lieutenant-captain"이라고 칭하였는데 이는 중대장(captain) 유고시 전투현장에서 중대장 자리를 유지하는 계급을 의미했다. 처음을 뜻하는 "first"가 계급에서는 처음을 뜻하지 않는다. 가장 먼저 다는 계급이 second lieutenant(소위)이다. 왜일까? 이는 중대장 유고시 가장 먼저(first) 현장을 유지하는(lieutenant) 계급이 소위보다 선임자인 중위(first lieutenant)이기 때문이다.

Quiz Answers & References

Quiz Answers

Chapter 1

A. MULTIPLE-CHOICE QUESTIONS

1. b
2. c
3. c
4. b

B. TRUE/FALSE QUESTIONS

1. F

(It also includes preserving peace, supporting policies, and implementing objectives.)

2. T
3. T
4. F

(All Soldiers, regardless of rank, are expected to live the Soldier's Creed and Army Values.)

C. FILL-IN-THE-BLANK QUESTIONS

1. Personal Courage
2. Landpower

Chapter 2

A. MULTIPLE-CHOICE QUESTIONS

1. b
2. b

B. TRUE/FALSE QUESTIONS

1. F

(A fire team has four Soldiers and is led by a noncommissioned officer.)

2. T

C. FILL-IN-THE-BLANK QUESTIONS

1. Fire team
2. Lieutenant

D. MATCHING QUESTIONS

1. d
2. a
3. c
4. b

E. SCENARIO-BASED QUESTIONS

1. b
2. b
3. c

Chapter 3

A. MULTIPLE-CHOICE QUESTIONS

1. c
2. b
3. a

B. TRUE/FALSE QUESTIONS

1. T
2. F

(It explains the preparatory command.)

3. T
4. F

(Neither of them is a command by itself.)

C. FILL-IN-THE-BLANK QUESTIONS

1. orderly manner, precision
2. preparatory command, command of execution
3. individual drill, subordinate unit drills

Chapter 4

A. MULTIPLE-CHOICE QUESTIONS

1. d.

(A balanced stance helps maintain accurate aim, control recoil, and indirectly supports faster, more effective shooting.)

2. b.

(The nonfiring hand stabilizes the rifle by supporting the forearm near the handguard.)

3. a.

(Keeping the trigger finger outside the trigger guard until ready to shoot ensures safety and prevents unintentional firing.)

4. a.

(A consistent stock weld helps maintain proper alignment between the sights and the target.)

B. TRUE/FALSE QUESTIONS

1. T

2. F

(The center of gravity is crucial for the shooter's stability during firing.)

3. T

4. F

(The grip should be firm but not too tight, not loose.)

5. T

6. F

(The butt stock should be positioned firmly in the shoulder to manage recoil and maintain stability.)

7. T

8. F

(Sight alignment is critical for accurate aiming.)

9. T

10. T

C. FILL-IN-THE-BLANK QUESTIONS

1. stance, balance
2. center of gravity
3. nonfiring, forearm
4. trigger, safety
5. butt stock
6. elbow

Chapter 5

A. MULTIPLE-CHOICE QUESTIONS

1. c.

(The shooter should focus on the front sight post because the human eye can only focus on one object at a time. Prioritizing the front sight ensures that it remains sharp and clear, allowing for accurate alignment with the target. The target and the rear sight aperture will appear slightly blurred, but this is intentional and necessary for precise shooting.)

2. b.

(A proper stock weld involves placing the cheek consistently on the stock in the same position for each shot. This consistent position ensures that the dominant eye aligns with the sights correctly, which is crucial for maintaining a stable and repeatable sight picture. Without a proper stock weld, variations in alignment can lead to inaccuracies.)

3. b.

(The sight picture refers to the visual image created when the shooter combines correct sight alignment <proper positioning of the front sight post within the rear sight aperture> with the placement of the aligned sights on the target. This combination ensures that the projectile follows the intended path to the target.)

4. c.

(At long ranges, gravity causes the bullet to drop below the point of aim. To compensate, the shooter must aim slightly higher than the intended point of impact. This adjustment, known as hold-over, accounts for the bullet's

trajectory and ensures it lands on the target. Simply focusing on the target or adjusting the stock weld will not correct for bullet drop.)

B. TRUE/FALSE QUESTIONS

1. F

(Sight alignment involves focusing on the front sight post, not the rear sight aperture.)

2. T

3. T

4. F

(The front sight should be in sharp focus, with the target and rear aperture slightly blurred.)

5. T

6. F

(The point of aim and the point of impact may not align perfectly at different distances due to factors like bullet drop and environmental conditions.)

7. F

(The instructor emphasized that accuracy is achieved by prioritizing focus on the front sight post, not the target.)

8. T

9. T

10. F

(Proper aiming requires focusing on the front sight while maintaining alignment with the target, not balancing focus between sights and the environment.)

C. FILL-IN-THE-BLANK QUESTIONS

1. stock weld, sight alignment

2. front sight post, blurred

3. sight picture, target

4. consistency, check position

5. point of aim, sight alignment

6. point of impact, bullet drop

7. front sight post, rear sight aperture

Chapter 6

A. MULTIPLE-CHOICE QUESTIONS

1. c

2. c

3. b

4. c

5. a

B. TRUE/FALSE QUESTIONS

1. F

(first aid must be applied by anyone at the point of injury.)

2. F

(MEDEVAC involves en route medical care provided to casualties during their evacuation.)

3. T

4. T

5. T

C. FILL-IN-THE-BLANK QUESTIONS

1. circulatory

2. tourniquet

3. hemorrhage

4. Tactical combat casualty care (TCCC)

5. tactical field care

D. MATCHING QUESTIONS

1. b
2. c
3. a
4. d
5. f
6. e

E. SCENARIO-BASED QUESTION

1. b
2. d
3. a

Chapter 7

A. MULTIPLE-CHOICE QUESTIONS

1. a
2. a
3. c (camouflage)
4. d (concealment)

B. TRUE/FALSE QUESTIONS

1. F
2. F
3. F
4. T
5. T
6. F
7. T
8. T
9. F
10. F (camouflage)

C. FILL-IN-THE-BLANK QUESTIONS

1. Cover
2. you, the enemy
3. direct, indirect
4. Concealment
5. shadows
6. Camouflage
7. natural, man-made

Chapter 8

A. MULTIPLE-CHOICE QUESTIONS

1. b
2. d
3. b
4. c

B. TRUE/FALSE QUESTIONS

1. F
2. T
3. T
4. F (360-degree)
5. T
6. F (rush)
7. T
8. T
9. T
10. F

C. FILL-IN-THE-BLANK QUESTIONS

1. low crawl, high crawl, rush.
2. low crawl
3. high crawl
4. rush

5. flexibility

6. arm-and-hand

7. prone position, next position, right leg, Rush

Chapter 9

A. MULTIPLE-CHOICE QUESTIONS

1. a
2. b
3. c
4. b
5. c
6. c

B. TRUE/FALSE QUESTIONS

1. F

(A platoon is composed of squads, which are made up of fire teams.)

2. T
3. T
4. F

(The platoon leader is responsible for accountability, leadership, and all the platoon does or fails to do.)

C. FILL-IN-THE-BLANK QUESTIONS

1. maneuver
2. four; nine
3. task; purpose
4. two

E. SCENARIO-BASED QUESTION

1. b

References

Chapter 1

FM 1, The Army, June 2005.

FM 3-0, Operations, October 2022.

ADP 1-01, Doctrine Primer, July 2019.

ADP 1, The Army, July 2019.

TRADOC Pamphlet 600-4, The Soldier's Blue Book, October 2023.

야전교범 0, 육군, 2019.

https://www.army.mil/

Chapter 2

https://www.defense.gov/

https://army.mil/

https://www.military.com/

FM 3-0, Operations, October 2022.

ADP 1-01, Doctrine Primer, July 2019.

TRADOC Pamphlet 600-4, The Soldier's Blue Book, 13 October 2023.

Chapter 3

TC 3-21.5, Drill and Ceremonies, May 2021.

TRADOC Pamphlet 600-4, The Soldier's Blue Book, October 2023.

Curriculum on Military Subjects, January 2019.

야전교범 참고-1-20, 육군, 2018

Chapter 4

p. 6-14, TC 3-22.9 (Rifle and Carbine), May 2016, Headquarters, Department of the Army

https://www.gungoddess.com/blogs/fundamentals/trigger-discipline-keep-your-finger-off-the-trigger

p. 6-4, TC 3-22.9 (Rifle and Carbine), May 2016, Headquarters, Department of the Army

p. 6-18, TC 3-22.9 (Rifle and Carbine), May 2016, Headquarters, Department of the Army

www.reedstargetshootingclub.co.uk

Chapter 5

https://thefiringline.com/forums/showthread.php?t=565759

http://mancavehq.weebly.com/merica/sight-alignment-sight-picture

p. B-3, TC 3-22.9 (Rifle and Carbine), May 2016, Headquarters, Department of the Army

https://www.atncorp.com/ballistic-calculator

Chapter 6

FM 4-02, Army Health System, November 2020.

ATP 3-21.8, The Infantry Rifle Platoon and Squad. 28 January 2024.

ATP 4-02.5, Casualty Care, May 2013.

TC 4-02.1, First Aid, January 2016.

Center For Army Lessons Learned, Tactical Combat Casualty Care Handbook, Version 5, May 2017.

https://www.naemt.org/

Chapter 7

www.army.mil

TC 3-21.75 (The Warrior Ethos and Soldier Combat Skills), Aug 2013, Headquarters, Department of the Army

ATP 3-21.8 (Infantry rifle platoon and squad), Jan 2024, Headquarters, Department of the Army

FM 3-90 (Tactics), May 2023, Headquarters, Department of the Army

JP 1-02 (Department of Defense Dictionary of Military and Associated Terms), Feb 2016, Department of Defense

국방일보

미 해군연구소 홈페이지

ARMA 게임 커뮤니티

Chapter 8

육군본부

TC 3-21.75 (The Warrior Ethos and Soldier Combat Skills), Aug 2013, Headquarters, Department of the Army

ATP 3-21.8 (Infantry rifle platoon and squad), Jan 2024, Headquarters, Department of the Army

FM 3-90 (Tactics), May 2023, Headquarters, Department of the Army

JP 1-02 (Department of Defense Dictionary of Military and Associated Terms), Feb 2016, Department of Defense

Chapter 9

FM 3-90, Tactics, May 2023.

FM 3-96, Brigade Combat Team, January 2021.

ADP 3-90, Offense and Defense, July 2019.

ATP 3-21.8, The Infantry Rifle Platoon and Squad. 28 January 2024.